Lectures on Mathematics in the Life Sciences

Volume 4

SOME MATHEMATICAL QUESTIONS IN BIOLOGY. III

The American Mathematical Society
Providence, Rhode Island
1972

Proceedings of the Fifth Symposium on
Mathematical Biology held in Chicago, December, 1970

edited by
Jack D. Cowan

International Standard Book Number 0-8218-1154-1
AMS 1970 Subject Classification 92A05

Prepared by the American Mathematical Society
with partial support from the National Science Foundation
under Grant Number GP 23303.

CONTENTS

FOREWORD

This volume contains lectures given at the fifth symposium on Some Mathematical Questions in Biology, held in Chicago on December 27, 1970, in conjunction with the annual meeting of the American Association for the Advancement of Science. The symposium was co-sponsored by the American Mathematical Society, and by the Society for Industrial and Applied Mathematics.

The lectures in this volume are concerned mainly with the topic introduced in the previous volume, the theoretical biology of development. The first two lectures by L. A. Segel and by A. D. J. Robertson respectively, deal mainly with the control of development in cellular slime molds, particularly with the chemically controlled aggregation of individual amoebae into the spatially organised system of such cells constituting the mold. The cellular slime molds are of considerable interest, both to theorists and to experimentalists, because they exhibit, in a relatively simple way, many of the cellular processes that are involved in the development of much more complex metazoons: growth, cell division, movement, differentiation, the formation of intercellular contacts, and secretion. Segel's lecture is concerned with a macroscopic approach to such problems, closely related to the work of A. M. Turing;[1] that of Robertson to the more microscopic approach introduced by B. C. Goodwin and M. H. Cohen.[2] The third lecture by B. C. Goodwin is an elaboration of this approach as applied to the fascinating problems of how the developing brain becomes differentiated in a highly specific fashion. The fourth and final lecture in this volume, by Elliot Montroll, is a reworking of one of the earliest attempts at biological theory, the Lotka-Volterra equations for ecological kinetics.[3] Montroll builds on the earlier work of E. H. Kerner[4] who first applied equilibrium statistical mechanics to such kinetics, and shows how many standard problems of quantum mechanics and noise theory have immediate application to ecological kinetics. Related applications of these techniques have been made to biochemical kinetics[5] and to neurodynamics.[6] As yet there is more mathematics in such theories than there is biology.

Taken together however, the lectures in this volume, and in the previous three, serve to indicate just how much of an interaction there has been between biology and mathematics. Let it continue!

It is a pleasure to acknowledge the support provided by the National Science Foundation.

Jack D. Cowan
Department of Theoretical Biology
University of Chicago
November 22, 1971

[1] A. M. Turing, Phil. Trans. Roy. Soc. **B**, **237** (1952), 37.

[2] B. C. Goodwin and M. H. Cohen, J. Theor. Biol. 25 (1969), 49-107.

[3] A. J. Lotka, *Elements of Mathematical Biology,* Dover, New York, 1956; V. Volterra, *Leçons sur la théorie mathématique de la lutte pour la vie,* Gauthier-Villars, Paris, 1931.

[4] E. H. Kerner, *A statistical mechanics of interacting biological species,* Bull. Math. Biophys. **19** (1957); *Further considerations on the statistical mechanics of biological associations,* Bull. Math. Biophys. **21** (1959); *On the Volterra-Lotka principle,* Bull. Math. Biophys. **23** (1961).

[5] B. C. Goodwin, *Temporal oscillations in cells,* Academic Press, New York, 1963.

[6] J. D. Cowan, in *Some Mathematical Questions in Biology,* **2**, ed. M. Gerstenhaber, Amer. Math. Soc., Providence, R. I., 1970.

ON COLLECTIVE MOTIONS OF CHEMOTACTIC CELLS

By

LEE A. SEGEL

Rensselaer Polytechnic Institute

1. Introduction and abstract

As an organism forms from a fertilized egg, cells multiply, specialize, migrate, and organize themselves. Of particular current interest are isolated collections of cells which exhibit such "cell choreography" in a relatively accessible manner. Mathematics may play an important role here and in other problems involving collective cell behavior. As more is being learned about how one cell interacts with another, there is increasing scope for deductive methods in ascertaining the behavior of assemblages of cells.

This paper summarizes some work I have done, mostly in collaboration with Evelyn F. Keller, on the collective behavior of cells whose motion is affected by the concentrations of certain chemicals. I find that understanding comes most easily by generalizing the answers to appropriately selected particular questions. Two such questions will be discussed here–why do certain amoebae suddenly aggregate into a number of collecting points or centers, and why do certain bacteria move steadily in bands down thin tubes?

This paper is written primarily for mathematicians with little training in biology but some interest in it. Hopefully, even biologists without much mathematical training will find the discussion of interest even though they may not be able to follow certain arguments in detail.

§2 contains the formulation and analysis of a relatively simple model for the sudden aggregation of certain amoebae. The onset of aggregation is viewed as an instability of a uniform distribution of amoebae and of a chemical called acrasin which is known to affect their

3

motion. A more complicated model is treated in §3 in order to determine more precisely the effects of an enzyme which destroys the effect of acrasin. The same fundamental flux equation as was used in the analysis of aggregation in §§2 and 3 is employed in §4 to analyze quite a different phenomenon, travelling nonlinear waves of bacteria. §5 employs a random walk model to clarify the mechanisms which lead to the fundamental flux equation. The paper concludes in §6 with some remarks concerning the role of mathematicians in biology.

2. Aggregation of amoebae: the simplest model

The cellular slime mold amoebae (Acrasiales) are normally found in soil and dung. Let us arbitrarily begin our discussion of their life cycle at the spore stage. A spore can be thought of as a dormant organism encapsulated in a protective shell. When conditions are favorable, the spore germinates or "hatches" and an active amoeba emerges. This one-celled organism moves about by extending portions of itself (pseudopods) and dragging itself after them. Ten to twenty microns in diameter, the amoebae "eat" by engulfing bacteria. When sufficient new matter has been generated by conversion of their bacterial food, the amoebae divide in two. Feeding and division continue as long as sufficient food is present.

When their food supply is exhausted, the amoebae seem to spread themselves uniformly over the space available to them. After an *interphase* period of a few hours, they begin to collect in a number of centers (Figure 1). The collected amoebae form into a worm-like slug but each amoeba retains its identity. The slug typically moves around for a time, then stops and erects a fruiting body composed of dead stalk cells bearing one or more clusters of spore cells (Figure 2). When conditions are again favorable, the spores germinate and the cycle repeats. For details of the life cycle, for other biological information,

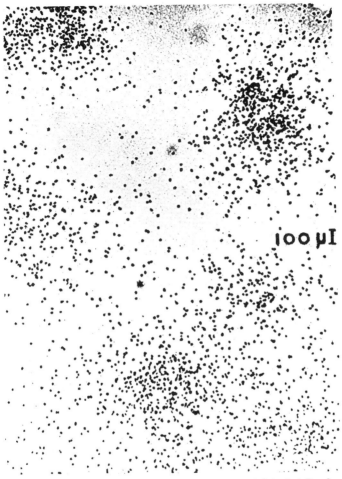

FIGURE 1. "Clouds" of amoebae formed at the beginning of the aggregation process. (From [56].)

and for an extensive bibliography, the reader is referred to the monograph of Bonner [1]. Robertson's paper [2] in the present volume provides much up-to-date information.

Biologists are interested in slime mold amoebae for two main reasons. Firstly, the aggregation phenomenon seems to be an excellent prototype of morphogenetic (form-producing) movement. "Purposeful" migration of cells from one place to another occurs frequently in the development of many species including man, but seemingly nowhere in a way less influenced by extraneous factors. Secondly, the production of the fruiting body is a simple example of (cellular) *differentiation*, which to a biologist means the splitting of a single cell type after some generations into two or more different cell types. In mammals, the original egg cell ultimately differentiates into a large number of different cells: nerve cells, heart cells, blood cells, etc.; with various subdivisions. In cellular slime mold, a fruiting body can result from the repeated division of a single amoeba but it contains two types of cells, stalk and spore.

Our theoretical work deals only with the first stages of aggregation. We now list some relevant biological facts, uncovered by the ingenuity and care of many experimenters. See [1] for information on classical experiments, and [3], [4], [5] for accounts of some important recent work.

(1) The feeding stage can be neglected in a discussion of aggregation, for aggregation will occur if a number of spores are induced to germinate in a foodless environment.

(2) In later stages the cells can become sticky and (in some species) pull each other toward the center in streams, but stickiness cannot play an essential role in the beginnings of aggregation, since aggregation can commence when the cells are widely scattered.

(3) The cells secrete a chemical called *acrasin* and they move preferentially toward relatively high concentrations of this chemical.

(4) If acrasin-water from attracting centers is pre-

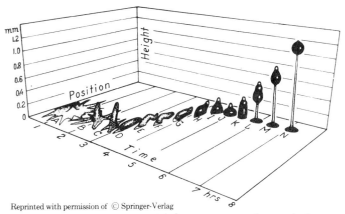

FIGURE 2. Behavior of an amoeba mass from the conclusion of aggregation to final formation of stalk topped by a fruiting body containing many spores. (From [57].)

sented to sensitive cells, they will be attracted to it, but not if the acrasin-water is held in a pipette for a few minutes before being presented to the cells. Thus the attractive activity of acrasin decays with time. It has been established that this decay requires the activity of an enzyme called *acrasinase.* (In general, the words "acrasin" and "acrasinase" apply to substances known to be present but not specifically identified.)

Culminating the efforts of many workers over a number of years is the recent indentification of acrasin, at least in some species, as a chemical called adenosine 3′, 5′-cyclic monophosphate or (cyclic) AMP for short. Identification of the chemical opens the way to the quantitative measurements called for by theories such as the one which we shall present. Furthermore, the particular identity which acrasin turns out to have gives additional reason to think that the mechanism involved in slime mold aggregation may have some degree of universality. This is because the chemical AMP is known to have important roles in mammalian physiology. Particularly striking is the fact that when hormones reach their target cells they release AMP within the cells, and this *universal* "second messenger" induces the special activity that the original *particular* first hormone messenger was to initiate.[1]

AMP has a molecular weight of 329. Its decay in activity is now known to be due to the bond-breaking activity of an enzyme. The enzyme, one thousand times as massive as the AMP, is a so-called "phosphodiesterase" and has been highly purified.

Even the bare outline of facts which we have presented makes manifest the necessity of providing a mathematical model in order to determine the consequences of the several interacting factors which influence aggregation. In Keller and Segel [6], the *onset* of aggregation in slime mold amoebae was characterized as an instability of the uniform distribution of amoebae and acrasin which seems to mark the interphase period. The instability is regarded as brought about by changes in individual amoeba properties, such as the acrasin secretion rate. The simplest plausible model which would exhibit such an instability was analyzed in the above-mentioned paper. This model will be treated now, while a more comprehensive model (and alternate approaches using the same formalism) will be discussed later.

Let $a(x, y)$ represent the density (number per unit area) of amoebae in the x-y plane. In the absence of birth and growth the requirement of amoeba conservation can be written

---

[1] See the *Scientific American* article by Bonner [46] for an interesting account of the identification of acrasin as cyclic AMP and the role of this chemical in hormone systems. See [47] for a more recent and more technical discussion of the importance of cyclic AMP, and its interaction with the calcium ion.

$$(2.1) \qquad\qquad \partial a / \partial t = - \nabla \cdot \boldsymbol{J}$$

where $\boldsymbol{J}$ is the amoeba flux density vector, the net number of amoebae per unit time which cross a unit line segment placed orthogonal to the direction of $\boldsymbol{J}$. We assume that $\boldsymbol{J}$ is the sum of two contributions, the first from a diffusive random motion and the second from a motion directed by chemical stimuli–in this case by relatively high concentrations of acrasin. Thus we write $\boldsymbol{J} = \boldsymbol{J}_{\mathrm{diff}} + \boldsymbol{J}_{\mathrm{chem}}$. As is well known, the random motion of molecules results on a macroscopic scale in a diffusion of heat or solute in which the flux can be regarded as proportional to the gradient in concentration of the diffusing substance. Analogously we assume here that

$$(2.2) \qquad\qquad \boldsymbol{J}_{\mathrm{diff}} = - \mu(\rho) \nabla a.$$

The proportionality factor μ is termed the *motility*. It measures the vigor of the random movement and might depend on the acrasin density ρ.

Motion directed by chemical concentration is called *chemotaxis*. (Greek $\tau \acute{\alpha} \xi \iota s$ = arrangement.) The classical work of Fraenkel and Gunn [7] gives a survey of many different kinds of animal "taxis" motions directed by chemical and other stimuli. A recent article [8] suggests that chemotaxis may be responsible for the accumulation of inflammatory cells at the site of virus-induced tissue destruction.

We postulate that the chemotactic flux density $\boldsymbol{J}_{\mathrm{chem}}$ is proportional to the gradient in chemical concentration: $\boldsymbol{J}_{\mathrm{chem}} \sim \nabla \rho$. Assuming no interference between amoebae, for a given chemical gradient twice as large a flux should result from the presence of twice as many amoebae. We thus write

$$(2.3) \qquad\qquad \boldsymbol{J}_{\mathrm{chem}} = a \chi(\rho) \nabla \rho$$

where the *chemotactic factor* χ measures the sensitivity of the cells to the chemical. The difference in signs in (2.2) and (2.3) arises from the fact that amoebae are expected to flow *away* from regions of high amoeba density and *towards* regions of high acrasin density.

In the language of rational mechanics, we have postulated a linear isotropic constitutive equation in assuming the additive nature of the diffusive and chemotactic contributions to the flux and the proportionality of each contribution to the gradient of the associated field. From this and the conservation equation (2.1) we obtain finally

$$(2.4) \qquad \partial a/\partial t = \nabla \cdot [\mu(\rho)\nabla a - a\chi(\rho)\nabla \rho].$$

Searching for the simplest equation which can describe the change in acrasin density ρ we postulate

$$(2.5) \qquad \partial \rho/\partial t = -k(\rho)\rho + af(\rho) + D\nabla^2\rho.$$

The first term on the right is responsible for a decay of acrasin activity with time. If k were a constant, this decay would be exponential, corresponding to a random deactivation of acrasin molecules. We know that the chemistry is not this simple. By letting k depend on ρ we can make the chemical description more realistic without, as will be seen, making our calculations more difficult.

The second term on the right in (2.5) represents the secretion of acrasin by the amoebae, at a rate $f(\rho)$ per amoeba. (There is some evidence that f depends on ρ.) The last term on the right represents ordinary diffusion, with diffusivity D.

Equations (2.4) and (2.5) have the uniform solution

$$(2.6) \qquad a = a_0, \qquad \rho = \rho_0,$$

providing that the constants a_0 and ρ_0 satisfy

$$(2.7) \qquad \rho_0 k(\rho_0) = a_0 f(\rho_0).$$

Let us examine the hypothesis that aggregation commences when the uniform solution becomes unstable to small perturbations. The stability analysis is straightforward [6]. One assumes that

$$(2.8) \qquad a = a_0 + \overline{a}, \qquad \rho = \rho_0 + \overline{\rho},$$

and finds equations for the perturbations $\overline{a}$ and $\overline{\rho}$ by substituting into the governing differential equations (2.4) and (2.5). Assuming small perturbations, one linearizes the equations, obtaining

$$(2.9a) \qquad \partial\overline{\rho}/\partial t = -\overline{k}\overline{\rho} + a_0 f' \overline{\rho} + f\overline{a} + D\nabla^2\overline{\rho},$$

$$(2.9b) \qquad \partial\overline{a}/\partial t = -a_0\chi\nabla^2\overline{\rho} + \mu\nabla^2\overline{a},$$

where a prime denotes a derivative,

$$\overline{k} = k(\rho_0) + \rho_0 k'(\rho_0),$$

and it is understood that in (2.9) and subsequent equations of this section, f, f', χ and μ are evaluated at the equilibrium acrasin level ρ_0. The equations admit solutions of the normal mode form

$$(2.10) \qquad \begin{aligned} \overline{a} &= \hat{a}\cos(q_1 x + q_2 y)\exp(\sigma t), \\ \overline{\rho} &= \hat{\rho}\cos(q_1 x + q_2 y)\exp(\sigma t), \end{aligned}$$

or

$$(2.11) \qquad \begin{aligned} \overline{a} &= \hat{a}\sin(q_1 x + q_2 y)\exp(\sigma t), \\ \overline{\rho} &= \hat{\rho}\sin(q_1 x + q_2 y)\exp(\sigma t), \end{aligned}$$

where $\hat{a}$, $\hat{\rho}$, q_1, q_2, and σ are constants. A wide class of initial conditions can be generated by Fourier synthesis using these solutions so, as is usual in formal stability calculations, one terms the basic solution (2.6) *stable* (*unstable*) to infinitesimal perturbations if the real part of the *growth rate* σ is negative (positive). Substitution of either (2.10) or (2.11) into (2.9) yields

$$(2.12) \quad \begin{aligned} (F - \sigma)\hat{\rho} + f\hat{a} &= 0, \\ a_0\chi q^2\hat{\rho} - (\mu q^2 + \sigma)\hat{a} &= 0, \end{aligned}$$

where

$$F = f'a_0 - \overline{k} - q^2 D$$

and $q \equiv (q_1^2 + q_2^2)^{1/2}$ is the wavenumber of the sinusoidal perturbations. The homogeneous equations of (2.12) possess a nontrivial solution for $\hat{a}$ and $\hat{\rho}$ if and only if σ satisfies a quadratic equation obtained by equating to zero the determinant of coefficients. The quadratic has real roots. These are both negative (stability) if and only if

$$(2.13) \qquad a_0(\chi f + \mu f') < \mu(\overline{k} + Dq^2).$$

The "most dangerous" perturbations are those with $q = 0$ (infinite wavelength) for then the right side of (2.13) is a minimum. From (2.13) with $q = 0$, then, we expect instability to occur (at least one positive root for σ) when

$$(2.14ab) \quad a_0\chi f + a_0\mu f' > \mu \overline{k} \quad \text{or} \quad a_0\chi f/\mu\overline{k} + a_0 f'/\overline{k} > 1.$$

Inevitable fluctuation of conditions will lead to the occasional appearance of locally high concentrations of amoebae and acrasin. Certain factors operate to disperse such concentrations while other factors operate to intensify them. Equation (2.14) quantitates the idea that instability will ensue if the latter outweigh the former.

To examine this more closely, consider the first term in (2.14b). The denominator contains stabilizing factors. The motility μ appears because a vigorous random motion leads to a rapid tendency for a local amoeba concentration to disperse, just as a large thermal conductivity leads to a rapid diffusive dispersion of a local hot spot.

The more rapid decay of chemical activity with high concentration is responsible for the stabilizing role of the effective decay constant $\bar{k}$.

The numerator of the first term in (2.14b) contains destabilizing factors. The appearance of χ reflects the fact that the bigger the chemotactic response of the cells, the more rapidly will the local concentration of acrasin attract more acrasin-producing cells. The factor $f = f(\rho_0)$ appears because a local concentration of cells which secrete acrasin at rate f will lead to a corresponding local concentration of acrasin.

There is evidence that the acrasin secretion rate increases with ambient acrasin concentration. Such an increase would enhance a local maximum of acrasin concentration, and this is reflected in the second term of (2.14b).

It is encouraging that the sensitivity of the amoebae to an acrasin gradient as well as the output of acrasin per cell both increase [9] about one-hundredfold [6] around the onset of aggregation. This marked increase of the destabilizing factors is the precise qualitative effect predicted by our theory. (It is absent in nonaggregating mutants.) Quantitative measurements to compare with (2.14) would provide a sterner test. These are made difficult by the extremely low levels of acrasin which are found naturally, but the discovery of a sensitive and accurate new assay for cyclic AMP [10] seems to make quantitative experiments feasible.

Another qualitative test of the theory is provided by experiments in which a few amoebae are confined to a very small area. It is easy to modify the analysis so that it is appropriate to the case where neither amoebae nor acrasin cross a circular boundary [6]. Aggregation is predicted to occur later, because larger values of $B \equiv \chi f + \mu f'$ are required, compared to the infinite case. The theory also makes the quantitative prediction that if aggregation takes place in two drops whose areas are small compared to the area of the usual aggregation "territory" then the corresponding values of B are in inverse ratio to the drop areas. If B increases linearly with time, the time at which aggregation commences should be a linear function of inverse drop area. Small drop experiments thus offer an excellent opportunity to test the quantitative predictions of the theory.

Konijn [11] performed experiments in which small drops of water and amoebae are placed on a special surface which somewhat mysteriously retains the amoebae within the original confines of the drop although the acrasin diffuses beyond the drop boundary. Aggregation occurs later in smaller drops as predicted by theory, but more quantitative tests would have to be done with drops placed on glass slides, for then (unlike Konijn's experiments which are specifically designed to the contrary) both amoebae *and* acrasin would stay within the drop boundary.

The perturbations assumed in (2.11) correspond to alternating straight bands of sparser and denser populations of amoebae. These seem never to occur but, as shown in Figure 3, the "clouds" of amoeba concentrations shown in Figure 1 can be modelled by superposing three bands at sixty degree angles, so that

(2.15)

$$a = \hat{a}[\cos qx + \cos(\tfrac{1}{2}qx - \tfrac{1}{2}\sqrt{3}qy) + \cos(\tfrac{1}{2}qx + \tfrac{1}{2}\sqrt{3}qy)].$$

Figure 4, taken from a paper of Gerisch [12], depicts a centrifugal aggregation which seems to be a mirror image of Figure 3. But one can obtain this pattern, which Gerisch termed "reticular," by making a correspondence in Figure 3 between hatched and sparser regions and nonhatched and denser regions. This is permissible since,

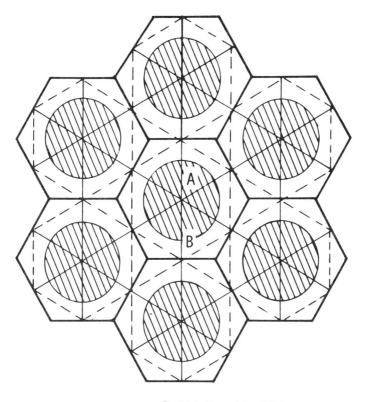

FIGURE 3. Superposition of sinusoidal bands of relatively high amoeba and acrasin density (light solid lines) to give hatched "cloud" pattern. Lines of relatively high density coincide at points like A. Two dotted lines of relatively low density meet a single high density line at B giving a below-average density. Full calculations of the pattern are as in [58]. (From [6].)

in a *linear* theory, the acrasin density a is only determined up to the arbitrary constant $\hat{a}$; in particular, the sign is undetermined.

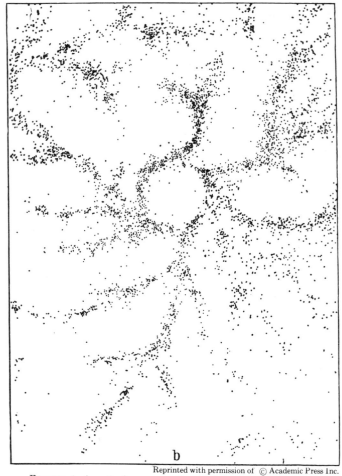

FIGURE 4. A reticular aggregation pattern. The cells move
outward in these experimental conditions so an attracting
center cannot be involved here. Note that the pattern appears
rather hexagonal in the center of the picture. (From [12].)

In previous years I have expended considerable effort
making various *nonlinear* stability analyses in order to
contribute to the understanding of hexagonal and band
patterns in so-called Bénard convection and a number

of related problems. (See for example [16] and [14].) It begins to appear that it would be worthwhile to perform such nonlinear calculations on slime mold stability problems, for these should show why it is that under some circumstances the aggregation pattern is cloud-like and in others reticular.

Another aspect of aggregation which should be illuminated by nonlinear calculations involves the triggering role of certain special cells. In some species (*Polysphondylium violaceum* [15] and *Dictyostelium minutum* [16], [17]) aggregation always seems initiated by special founder cells which act as centers of attraction. In *D. discoideum,* however, *centerless* patterns appear when pains are taken to obtain a homogeneous cell population [12]. By contrast, aggregations of more natural configurations (with centripetal radial streams) appear in homogeneous cell populations of *D. discoideum* when multi-cellular agglutinates of greater maturity are introduced and act as artificial centers.

It is helpful here to invoke a metaphor. Think of the special cells as a trigger and the other cells as the remainder of a pistol. Under certain circumstances, the gun will fire only when the trigger is pulled (agglutinates or founders trigger aggregation). Under other circumstances the gun is so delicately cocked that a footfall will set it off (random perturbations initiate aggregation in sensitive homogeneous populations). The latter case is covered by linear stability theory with its study of "infinitesimal" perturbations. The former case requires an analysis of the system response to finite amplitude perturbations, and this is the subject of nonlinear stability theory. Nonlinear calculations are planned, and it appears likely that they will harmonize some observations of aggregation in *D. discoideum.*

The question arises, is it adequate to assume, as we

have above, that the acrasin output is simply a function f of the local acrasin level r? For it happens that certain species, including $D.$ $discoideum$ [18] and $P.$ $violaceum$ [19] seem to aggregate in a rhythmically pulsing fashion, at least some of the time. Cohen and Robertson [20], [21] have enlarged on observations and conjectures concerning the pulsing, particularly those of Shaffer, to erect a competitive theory of aggregation based on the hypothesis that secretion of acrasin in a pulse τ_p seconds long is triggered when the ambient acrasin level exceeds a certain threshold T. (We have an alternative explanation for rhythmic aggregation-see §3.) There is presumed to be a delay of τ_d seconds between signal and secretion. A refractory period of τ_r seconds in which no acrasin can be secreted is assumed to follow each pulse.

For details, the reader is urged to consult the paper of Robertson in this volume. He will note that the approach of Cohen and Robertson is a "microscopic" one dealing with the behavior of single cells in contrast to our "macroscopic" approach in which the cell density is treated as a continuous variable. The more "exact" microscopic model leads to a much less tractable mathematical problem so the *analysis* (not the model) of Cohen and Robertson inevitably contains more approximations than Keller's and mine. This is a secondary matter; the central question is whether or not acrasin excretion is significantly pulsatile. Even if it is, I believe that analysis of a continuum model is the best way to obtain a clear understanding of system response. However, the hypothesis that f, the acrasin secretion per unit amoeba density at point x and time t, is simply a function of $\rho(x,t)$ must now be modified. Assuming for simplicity that the pulse is at the constant level K one would have to consider a history-dependent secretion rate given as follows:

(2.16)

$$f(t) = 0 \qquad \text{if } f(t') > 0 \qquad \text{when } t - \tau_r \leqq t' \leqq t,$$

$$\text{or if } \rho(t') \leqq T \qquad \text{when } t - \tau_d - \tau_p \leqq t' \leqq t - \tau_d.$$

$$f(t) = K \qquad \text{if } f(t') = 0 \qquad \text{when } t - \tau_r \leqq t' \leqq t,$$

$$\text{and if } \rho(t') > T \qquad \text{when } t - \tau_d - \tau_p \leqq t' \leqq t - \tau_d.$$

If significant cell motion typically took place during the delay period τ_d, a Lagrangian or material description would have to be adopted. In this case an initial particle position X would be implicit in (2.16) so (2.16) would refer to events at a given particle. Otherwise x would be implicit in (2.16), which would refer to events as a given *position*. It is apparent that the problem is far more difficult if a cell's previous history significantly influences its current acrasin output.

I feel that the occurrence of reticular centerless aggregation in homogeneous assemblages of *D. discoideum* provides grounds for believing that history dependence is negligible in that species at least under some circumstances. (A "small" degree of history dependence disappears in a macroscopic model, with its time-and space-averaged viewpoint.)

Cell stickiness is another factor which may be important under some circumstances. It certainly cannot influence the beginnings of aggregation when the cells are not in contact. It is not important even in later stages of reticular pattern formation in homogeneous assemblages of *D. discoideum* [12, p. 695].

There may be a threshold value of acrasin gradient below which no chemotactic response occurs. The existence of such a threshold can be taken into account crudely in a very simple manner [6] (see below) but a more sophisticated calculation would involve retaining the $a_0 \chi \nabla^2 \bar{\rho}$ term in (2.9b) if and only if $|\nabla \bar{\rho}| > R$, for some constant threshold R. Quite an interesting mathematical problem emerges, for the stability analysis now requires consideration of a piecewise linear system. Superposition is no longer possible, and the location of the curves bounding regions where the gradients are too small to induce chemotaxis is unknown. The existence of the threshold R is conjectured on the grounds of general biological intuition, not because of evidence in this particular context. For this reason, the biological case for pursuing an investigation of the effect of threshold on stability is not compelling, but I think that the problem has sufficient mathematical interest to warrant investigation.

In concluding this section, it is worth recording the prediction of our model in possible instances where acrasinase is absent, so $k \equiv 0$. (This appears to be the case in the *Polysphondelium* species.) From (2.7), for a uniform solution we must have no acrasin output $[f \equiv 0]$. Since bacteria are known to secrete cyclic AMP [22] it has been supposed that some degree of sensitivity $(\chi > 0)$ is useful in finding food. Positive sensitivity in itself brings forth no consequences even in the absence of acrasinase. But from (2.14) the condition for instability is now $f\chi > 0$, so with positive χ aggregation is predicted to commence as soon as the cells begin to secrete acrasin.

Suppose that there is some acrasin output during the interphase period. Then the acrasin level, although spatially uniform, would increase with time prior to the onset of aggregation. For good results, a new stability analysis of a time-varying solution would have to be performed, but we can get a rough idea of the requirements for instability using (2.14a) with $k = 0$ and with a "typical" value of the acrasin level selected as ρ_0 in the various auxiliary formulas. According to (2.14a) instability will set in when $\chi f > \mu(-f')$, which leads to the suggestion that if enzyme is absent, *inhibition* of acrasin production by a relatively high ambient acrasin level ($f' < 0$) will be present to prevent premature aggregation. (Remember, dispersal is helpful in obtaining food; aggregation is advantageous only when the food supply is exhausted.) The presence of acrasinase, then, acts to keep aggregation from commencing too soon, and allows a positive feedback from acrasin level to acrasin production ($f' > 0$). In the next section, we shall examine the action of acrasinase in more detail.

3. A more accurate model for aggregation

Biologists speak of the "territory" associated with a given center of aggregation, by which they mean the area from which the center draws its cells. Territory size has been measured in a number of circumstances. The most striking result here is that under certain environmental conditions, the territory size seems almost independent of cell density. It is natural to associate territory size with the critical wavelength, i.e., the wavelength of the so-called "most dangerous" disturbance modes which are the first to grow when instability commences. But as mentioned in our discussion of (2.13) the critical wavelength is infinite according to the model of §2.

It is possible to obtain finite territory size predictions from our analysis (i) by including threshold effects, (ii) by identifying territory size with the wavelength of the fastest-growing disturbance, (iii) by considering a more sophisticated model. Let us consider these possibilities in turn.

As indicated in [6], threshold effects can be crudely estimated by assuming that chemotaxis is absent unless the spatial average of the relative acrasin gradient $|\nabla \rho/\rho|$ is larger than some threshold α. This is easily seen to require that the wavelength λ not exceed an upper bound $\lambda_u \equiv 2\pi\hat{\rho}/(\alpha\rho_0)$. There is a concomitant addition of a stabilizing positive term to the right side of (2.14b). Instability will set in when this stabilizing effect is least, which occurs when λ takes on its upper bound. Thus this approach predicts a territory size λ_u. As expected, according to this point of view, territory size will depend on the amplitude of the disturbance.

When conditions for disturbance growth are first met, the disturbance wavelength of the fastest growing disturbance is identical with the critical wavelength and so is infinite. As sensitivity and acrasin output continue to increase the wavelength $\lambda_{max} = 2\pi/q_{max}$ of the fastest growing disturbance is finite and is easily calculable by finding the wavenumber q_{max} which maximizes the growth rate σ. The result is

$$(3.1) \qquad \lambda_{max} = 2\pi \left(\frac{2D}{\bar{k}\Delta}\right)^{1/2} \left[\, 1 + O(\Delta)\, \right]$$

where, from (2.14a),

$$\Delta \equiv a_0(\chi f + \mu f')/\mu\bar{k} - 1$$

is a measure of how far critical conditions are exceeded.

For λ_{max} to be determinable by experiment, the acrasin secretion rate and the sensitivity would have to approach well-defined limiting values. Even if this were the case, predictions of the theory should be regarded with caution, for the growth of perturbations casts doubt on the validity of linearization.

Stability theory contains several examples where an infinite critical wavelength is replaced by a finite one when a more accurate model is analyzed, so possibility (iii) is certainly worth investigating. A more accurate model would take into account the fact that the decay of acrasin activity is caused by the bond-breaking action of the enzyme acrasinase. A simple but plausible chemical reaction scheme would have the enzyme and acrasin (concentrations η and ρ) temporarily uniting to form a complex (concentration c) which would dissociate into free enzyme and de-activated acrasin. In chemical symbolism

$$(3.2) \quad \rho + \eta \underset{k_{-1}}{\overset{k_1}{\rightleftarrows}} c \overset{k_2}{\rightarrow} \eta + \text{de-activated acrasin},$$

where the k's are the rate constants for the various steps. Substantially as in [6] the governing system would now be

$$(3.3) \quad \begin{aligned} \partial a/\partial t &= \nabla \cdot [\mu(\rho) \nabla a - a\chi(\rho) \nabla \rho], \\ \partial \rho/\partial t &= -k_1\rho\eta + k_{-1}c + af(\rho) + D\nabla^2\rho, \\ \partial c/\partial t &= k_1\rho\eta - (k_{-1} + k_2)c + D_c\nabla^2 c, \\ \partial \eta/\partial t &= -s(\eta)\eta - k_1\rho\eta + (k_{-1} + k_2)c \\ &\quad + ag(\rho, \eta) + D_\eta\nabla^2\eta. \end{aligned}$$

Here g is the rate of enzyme secretion, D_η and D_c are diffusion constants for the enzyme and the complex, and s is the rate at which the enzyme concentration decays.

These equations have the uniform solution

$$(3.4) \quad a = a_0, \quad \rho = \rho_0, \quad c = c_0, \quad \eta = \eta_0$$

providing

$$(3.5) \quad \begin{aligned} k_1\rho_0\eta_0 &= k_{-1}c_0 + a_0f(\rho_0), \quad k_1\rho_0\eta_0 = (k_{-1} + k_2)c_0, \\ s(\eta_0)\eta_0 &= a_0g(\rho_0, \eta_0). \end{aligned}$$

Isolated enzyme seems to retain its activity for long periods, so s is very small, perhaps effectively zero. If we assume $s = 0$ a steady solution requires $g(\rho_0, \eta_0) = 0$. Should uniformly active enzyme be continually produced, the interphase solution would be spatially uniform but would vary with time, and a much more difficult stability analysis would be required. Measurements of enzyme production are not accurate enough at present to guide the choice of g.[3]

Beth Stoeckly and I have been working on the stability of the solution (3.4). The work is not completed, but there are some interesting results worth summarizing here. We write

$$(3.6) \quad \begin{bmatrix} a \\ \rho \\ c \\ \eta \end{bmatrix} = \begin{bmatrix} a_0 \\ \rho_0 \\ c_0 \\ \eta_0 \end{bmatrix} + \begin{bmatrix} \hat{a} \\ \hat{\rho} \\ \hat{c} \\ \hat{\eta} \end{bmatrix} \begin{matrix} \cos(q_1 x + q_2 y) \\ \text{or} \\ \sin(q_1 x + q_2 y) \end{matrix} \quad e^{\sigma t}$$

and find upon substitution and linearization that σ satisfies a quartic of the form

$$(3.7) \quad \begin{aligned} \epsilon\beta\sigma^4 &+ (\lambda_1 q^2 + \lambda_0)\sigma^3 \\ &+ (\theta_2 q^4 + \theta_1 q^2 + \theta_0)\sigma^2 \\ &+ (\nu_3 q^6 + \nu_2 q^4 + \nu_1 q^2 + \nu_0)\sigma \\ &+ (\Omega_4 q^8 + \Omega_3 q^6 + \Omega_2 q^4 + \Omega_1 q^2) = 0, \end{aligned}$$

where

$$(3.8) \quad \epsilon \equiv c_0/\rho_0, \quad \beta \equiv \eta_0/\rho_0, \quad q^2 \equiv q_1^2 + q_2^2,$$

and where the coefficients λ_i, θ_i, ν_i and Ω_i depend on complicated combinations of the various parameters of the problem. The growth rate $\sigma \equiv \sigma_r + i\sigma_i$ can now be complex. The marginal state $\sigma_r = 0$ which divides stability from instability can occur when $\sigma_i = 0$ (monotonic in-

---

[3] As this paper is being completed, I am informed by Dr. Gerisch of evidence that the cells of *D. discoideum* secrete an inhibitor of acrasinase. The presence of the inhibitor can be adequately modelled in some cases by allowing the decay coefficient s to depend on the acrasinase concentration η. Its simplest effect is to reduce the equilibrium (acrasinase) concentration η_0 and hence, by (3.9), to destabilize by decreasing $\bar{k}$ in (2.14).

stability) or when $\sigma_i \neq 0$ (oscillatory instability). We must maximize σ_r as a function of q^2 to find the value of the critical wavenumber.

To set out all possible cases is a forbidding task, and an unprofitable one. In the absence of information concerning the parameters, the most helpful thing to do is to point out new qualitative possibilities inherent in this model. This can be done most efficiently by making various plausible simplifying assumptions about the magnitudes of the various parameters. Here are some of the results. In the ensuing equations, all functions should be understood as evaluated at equilibrium conditions. Subscripts on g denote partial derivatives.

Case 1. We assume

$$\epsilon \ll 1, \quad g = g_\rho = g_\eta = s = 0,$$

and either

$$\beta = O(1) \quad \text{or} \quad \beta \ll 1, \ \beta/\epsilon = O(1).$$

We predict monotonic instability at infinite wavelength when (2.14) is satisfied if

$$(3.9) \quad k(\rho) = \eta_0 k_2 K/(1 + k_3\rho), \quad K = k_1/(k_{-1} + k_2).$$

This bears out a suggestion of Keller and Segel [6] that the assumptions they made in passing from (3.3) to the simplified system (12.4)–(12.5) could be formally justified by singular perturbation arguments in the spirit of the discussions of chemical kinetics by Bowen, Acrivos and Oppenheim [23] and by Heineken, Tsuchiya and Aris [24].

Case 2. We assume

$$\epsilon \ll 1, \quad g = g_\eta = s = 0, \quad g_\rho > 0.$$

If the instability is monotonic, it occurs at the finite wavelength

$$(3.10) \qquad 2\pi \left[\frac{2\mu D\rho_0}{a_0\mu(\rho_0 f' - f) + a_0\rho_0\chi f} \right]^{1/2}.$$

Oscillatory instability can occur in principle. Full treatment of this possibility requires analysis of a cubic with complicated coefficients, but it is easy to show that when $\beta f' < f$ overstability will not occur for small q (and instabilities never occur for large q). Thus when $\rho_0 f' < f$, even if instability does commence in an oscillatory fashion, it will be with finite territory size.

Case 3. We assume

$$\epsilon \ll 1, \quad \beta \ll 1, \quad \beta/\epsilon = O(1), \quad g = g_\eta = s = 0, \quad g_\rho > 0,$$

and find that the first disturbance to grow will be oscillatory. This overstability will set in (at infinite wavelength) when

(3.11)

$$\frac{a_0^2 f g_\rho}{c_0\rho_0 k_1} + a_0\gamma^2 f' > \frac{\eta_0\gamma f}{c_0} + a_0\gamma g_\rho, \quad \text{where} \quad \gamma \equiv 1 + \frac{\eta_0}{c_0}.$$

The frequency of oscillation is given by

$$(3.12) \qquad a_0 \left[\frac{g_\rho f}{c_0 + \eta_0} \right]^{1/2}.$$

It is not appropriate here to comment extensively on the biological significance of these results, so we shall limit ourselves to the following remarks. (1) We have demonstrated that the interaction of amoebae, acrasin, and acrasinase can under certain conditions give rise to an instability which commences with a well-defined finite wavelength. (2) We have shown that overstability can occur, which means that aggregation can commence with a pulsing contraction of the whole territory. This type of pulsation is not mentioned in the literature, but Konijn (private communication) reports that *D*.

rosareum, a newly discovered species, exhibits a phenomenon which seems of the kind to which the present analysis would apply. Aggregation in *D. rosareum* involves overall pulses separated by very long time intervals (about twenty minutes) during which the cell pattern almost disintegrates. (3) The principal assumption we made in deriving our results is of the common Haldane type wherein the concentration of chemical substrate (acrasin) is large compared to the concentration of complex and/or enzyme. It is not known, however, whether this assumption is applicable in the present context. If not, as yet unexplored portions of parameter space will have to be examined with the aid of a computer. But in any case, our analysis points to the type of qualitative results which can be anticipated. (4) Our most interesting results are subject to the assumption that $g_\rho > 0$, i.e., that the acrasinase output is stimulated by relatively high levels of acrasin. Such an assumption is biologically plausible.[4] That it leads to oscillations is suggested by the argument more acrasin $\Rightarrow$ more enzyme $\Rightarrow$ less acrasin $\Rightarrow$ less enzyme $\Rightarrow$ more acrasin $\Rightarrow \cdots$. This argument presumes that the instability is essentially chemical–which is the case, since neither the frequency of (3.12) nor the instability condition (3.11) contains the coefficients μ and χ or the diffusivities D, D_η and D_c. To a first approximation, the cells merely follow the chemical concentration.

Interesting extensions of the present research can be

---

[4] Coupled with the assumption $g = g(\rho_0, \eta_0) = 0$, $g_\rho(\rho_0, \eta_0) > 0$ means that enzyme is effectively absorbed when ρ falls below ρ_0. Actual absorption seems unlikely, but effective absorption could be caused by an inhibitor. Further, if $g(\rho_0, \eta_0)$ were assumed positive the problem of acrasinase absorption would not arise. The analysis would become difficult to perform analytically in this case, but it seems reasonable to suppose that oscillations would still occur.

made in a number of directions. Several possibilities have
been mentioned. Another has already been accomplished
by Edelstein [25] in his analysis of the sorting of two-
cell species with different chemotactic behavior. To
mention one more, further investigations of oscillatory
aggregation would be of value. Our linearized analysis has
revealed the possibility of certain kinds of oscillations,
but nonlinearities would have to be taken into account if
one were to have a chance of using the present model to
explain the fact that a number of waves can occur in a
single aggregation territory. If these oscillations can not
be accounted for by an extension of the present theory,
positing the internal delay which form the essential
assumption in the papers of Cohen and Robertson is indeed essential for complete understanding of patterns
in aggregating slime mold amoebae. But the alternative
remains that all major phenomena can be accounted for
by an analysis of the essentially instantaneous response
of the amoebae to a collection of reacting chemicals.
Depending on the circumstances, one should analyze the
periodic disturbances considered in nonlinear stability
theory or the effect on the system of an isolated source
of acrasin. Such calculations are planned, and should
appreciably extend the range of phenomena covered by
the theory.

4. Travelling bands

A major factor in the striking successes of modern
biology has been the realization that the behavior of
micro-organisms constitutes an *accessible model* for the
behavior of higher organisms. The *accessibility* resides
in two facts: (1) Micro-organisms offer examples of given
phenomena which (in many aspects) are far simpler than
those of higher organisms, so much so that detailed
biochemical and genetic analysis has often proved
possible. (2) Micro-organisms reproduce rapidly (a gen-

eration every 20 minutes is typical for bacteria) and therefore offer great chance for the appearance of mutants which perform a given biochemical task in various alternative and usually defective ways. By combining mutants or by supplying them with various intermediate products of a reaction sequence, biologists can discover with amazing precision many facts about the normal course of biochemical processes, and of the genetic processes which direct them. That micro-organisms are a *model* seems to follow from what can be regarded as the fundamental simplifying feature of biology, the existence of evolution. Higher organisms evolve from lower, so that one can expect behavior of the former to be founded on combined and elaborated versions of the latter's mechanisms.

Of considerable current interest to a number of biologists is the problem of sensory transduction. How does an organism receive a signal and translate it into appropriate action? M. Delbruck and his collaborators have chosen as a model the fungus *Phycomycetes* whose growth rate alters in response to such stimuli as light, gravity, and stretch [26]. J. Adler has selected as his model chemotactic strains of the common intestinal bacteria *Escherichia coli* [27], [28]. *E. coli* has served as a model in countless other investigations, so more information is probably available about this bacterial species than about any other. This alone confers a great advantage on the investigator who uses *E. coli*.

The investigations of Adler and his associates take advantage of the fact that certain strains of *E. coli* are motile (capable of movement) by virtue of one or more whip-like flagellae. If these bacteria sense a relatively low concentration of oxygen or of their "fuel supply" then they apparently tend to change direction and thus to move towards regions of higher concentration of these chemicals. The moving bacteria congregate in bands visible to the naked eye. Changes in the movements of individual bacteria thus provide tangible (or, more precisely, ocular) evidence that the bacteria are receiving external stimuli. It is worth emphasizing that it is not feasible to dispose a few molecules of chemical in front of a single bacterium and to observe its response. Observations are necessarily of bacterial *populations* confronted with varying concentrations of chemical.

Keller and Segel [29] showed that mathematical analysis can unify various experimental observations concerning chemotactic bands of bacterial populations. The equation for cell flux is the same as that used in the analysis of slime mold aggregation. Confidence in its applicability is increased by the seemingly successful use of the same formalism in the study of two rather different systems. This is perhaps sufficient justification for trying to analyze Adler's experiments. In contrast with developmental biology, however, the macroscopic events are here not of great interest in themselves. Nevertheless, a precise and unifying theory of these events should be of value in uncovering the biochemical mechanisms which underlie them.

A typical experiment of the class which we wish to analyze is the following [27]. A thin glass tube is filled with a solution containing known chemicals, principally an energy source such as serine. One end of the tube is sealed and a number of motile bacteria are introduced into the other. Then the other end of the tube is sealed. Soon a band of bacteria can be seen with the naked eye moving steadily down the tube. Many bacteria are still left where they were placed, but after a time a second band follows the first. The bacteria which remain at the end of the tube are no longer motile, but this is not because of any irreversible defects. If these bacteria are placed in fresh tubes they too will form bands.

Analysis of chemical composition as a function of distance down the tube makes it clear that the bacteria typically consume all of the oxygen in their vicinity (and some of the serine). "Avoidance" of low oxygen concentration induces a number of bacteria to move down the tube, consuming all the oxygen as they go in order to utilize the serine fuel. The second band appears because in the absence of oxygen the bacteria switch over to an *anerobic* (without oxygen) system of utilizing fuel, and thus consume the remaining serine.

We shall assume that only gradients in a single *critical chemical* significantly affect the motion of the cells. As above we shall employ equation (2.4) to describe the change in cell density. Using x to denote distance along the tube, and assuming that the only significant spatial variations are with respect to x, we write

$$(4.1) \qquad \frac{\partial b}{\partial t} = \frac{\partial}{\partial x} \left[\mu(s) \frac{\partial b}{\partial x} - b\chi(s) \frac{\partial s}{\partial x} \right].$$

Here b (replacing amoeba density a) represents the density of bacteria while s (replacing acrasin density ρ) stands for the concentration of the critical chemical.

In this problem it is necessary to specify the functions μ and χ. We shall assume

$$(4.2) \qquad \mu = \text{constant}; \quad \chi = \delta/s, \ \delta \text{ a constant}.$$

There is no experimental information available, so these assumptions are in essence informed guesses. The assumption for χ seems in accord with the Weber-Fechner "law" wherein response is proportional to relative stimulus ($J_{\text{chem}} \sim \nabla s/s$) but the most compelling argument in its favor is that with constant μ the chosen form for χ is the least singular which will allow a travelling band solution to the problem. For a proof and for further discussion, see [29].

There is evidence that the rate of chemical consumption, even at relatively low concentrations, is limited mainly by the ability of the bacteria to consume it. We thus assume that

(4.3) $\partial s/\partial t = -kb$, k a constant.

We have neglected diffusion of the chemical. Using the solution to be obtained below, we can show that the neglected term is small compared to those retained. (Also, a numerical analysis done by the author and T. Scribner shows that the results are essentially unchanged by the introduction of chemical diffusion.)

As the reader will doubtless have anticipated, we identify "travelling bands" with "travelling waves" and look for a solution of the form

(4.4) $s = s(\xi)$, $b = b(\xi)$, $\xi = x - ct$, c a constant.

We assume in addition that far in front of the band both the density and the flux of bacteria vanish, while the concentration of chemical approaches a constant s_∞ (the initial chemical concentration). Thus

(4.5) $b \to 0$, $db/d\xi \to 0$, $s \to s_\infty$ as $\xi \to \infty$.

The problem formed by (4.1)–(4.5) can be solved. Choosing the origin of coordinates in order to obtain relatively simple formulas we make the abbreviations

$$\bar{\xi} = c\xi/\mu, \qquad \bar{\delta} = \delta/\mu,$$

and find

(4.6a,b)

$$\frac{s}{s_\infty} = \left[1 + e^{-\bar{\xi}}\right]^{-1/(\bar{\delta}-1)},$$

$$\frac{b}{c^2 s_\infty (\mu k)^{-1}} = \frac{1}{\bar{\delta}-1} e^{-\bar{\xi}} (1 + e^{-\bar{\xi}})^{-\bar{\delta}/(\bar{\delta}-1)}.$$

The useful relation

(4.7) $c = Nk/(As_\infty)$

can be obtained by integrating (4.6b) or, more easily, by integrating the equation (4.3). Here

$$N = A \int_{-\infty}^{\infty} a(\xi)\, d\xi$$

is the number of bacteria in the band and A is the cross-sectional area of the tube.

It is easily seen that s and a remain finite as $\xi \to -\infty$ if and only if $\bar{\delta} > 1$. Thus for travelling bands to exist it is necessary that $\bar{\delta} > 1$, i.e. for $\delta > \mu$. (The disordering effect of random motion must be outweighed by the organizing effect of chemotaxis.) If $\bar{\delta} > 1$, bacterial density and chemical concentration decay rapidly behind the band to a value which is effectively zero. This is in agreement with observation.

From (4.6), then, we see that when $\bar{\delta} > 1$ the equations admit a solution in the form of a steadily travelling band for every value of the wavespeed c. The situation can also be regarded as one in which, given any positive number N, one can obtain a unique band which contains N bacteria and travels steadily at speed $c(N)$ given by (4.7).

Comparison with experiment is afforded by (4.7), for data is available for every parameter therein. This data is rather imprecise, however, particularly in the case of N. Agreement between theory and experiment is about as close as can be expected. In two cases the predicted values of wavespeed c are (in units cm/hr) 1.5 and 1.2 while corresponding observed values are 0.9 and 2.0.

More convincing evidence for the worth of the theory is provided by qualitative aspects of the bands. Adler observed that densitometer plots of bacterial density *vs.* distance were sometimes steeper toward the front of the band, and sometimes toward the rear. The same features are predicted by (4.6), depending on whether

$\bar\delta > 2$ or $1 < \bar\delta < 2$. The ratio of chemotactic strength to motility, $\bar\delta$, can be estimated by comparing observed and predicted values of the band width. In one case $\bar\delta$ was estimated to be between 1 and 2; in another $\bar\delta$ was estimated to have the value 3. As predicted, the trailing edge of the band was steeper in the former case, the leading edge in the latter. See [29] for details of the comparison between theory and experiment.

There is rough quantitative agreement with experiment and good qualitative agreement. It appears that the theory deserves further study. Most important are measurements of the dependence of χ and μ on chemical concentration. Such measurements are attempted in the paper of Adler and Dahl [30] wherein the idea of associating random motion with an effective diffusion constant or motility (μ in our notation) was introduced. Adler and Dahl found that bands did not form if the chemical methionine was omitted from the solution with which they filled the tubes. They therefore used ordinary diffusion theory to measure μ, obtaining values of the order of $\frac{1}{4}$ cm^2/hr. However their curves of distance "diffused" $vs.$ time did not fit the predictions of diffusion theory very well. The present analysis suggests that chemotaxis was present but that $0 < \bar\delta < 1$. Some further calculations now in progress should open the way to more accurate determination of μ along the lines of the Adler-Dahl experiments.

Additional calculations are also needed to quantitate the latest assay procedure of Adler [28] in which the capillary tube is inserted for an hour into a solution containing a number of chemotactic bacteria. (The greater the degree of chemotaxis, the larger will be the number of bacteria found in the tube.) Somewhat similar calculations are required to quantitate the experiments of Smith and Doetsch [31].

Initial value problems of a rather obvious nature are appropriate to the contexts just mentioned. Sufficiently accurate solutions should be obtainable without difficulty by fairly standard techniques. Much more difficult is the task of showing that suitable initial value problems lead eventually to solutions which have the form of steadily travelling bands. This will require determining how the initial conditions influence the number N of bacteria which enter the band.

The problem is somewhat similar to, but more difficult than, the flame propagation study of Gel'fand [32] wherein a one-parameter of travelling waves was also found and where an elegant analysis of the initial value problem showed that only the wave of fastest speed would occur naturally. Other distinguished Soviet mathematicians have made contributions in this area, for Kolmogoroff, Petrovsky, and Piscounoff [33] studied the emergence of a travelling wave solution to $\partial v/\partial t = \nabla^2 v + F(v)$, and they discussed certain biological applications of this problem.

The question here, to put it dramatically, is this: by what rule is it decided how many individuals will join the band who travel steadily off in search of favorable conditions and how many will remain where they were placed, immobile and doomed? The matter is under investigation.

5. A random walk model for chemotaxis

Our previous discussion has shown that a phenomenological macroscopic viewpoint is useful when investigating the collective behavior of cells. One reason for this utility is that considerable progress can be made in analyzing macroscopic phenomena without waiting for as yet undiscovered microscopic information. A more fundamental reason is that understanding the response of a collection of cells would require further analysis even if complete knowledge were available concerning every biochemical mechanism in an *individual* cell's interior.

Nevertheless, it should be helpful to have some understanding of how events at the microscopic level can act to produce phenomena at the macroscopic level. (Throughout this paper, "microscopic" is synonymous with "cellular" while "macroscopic" refers to collective cell behavior.) Such an understanding is particularly to be desired in view of the fact that the cells exhibit a response to chemical gradients which might seem virtually imperceptible to such a small organism. This is particularly true for chemotaxis in bacteria, for bacteria are an order of magnitude smaller than slime mold amoebae. Consequently Keller and Segel [34] presented a random walk model for chemotaxis that shows how the erratic actions of single cells, which respond to locally sensed chemical concentrations, can result in an average cell flux which is proportional to the concentration gradient. We shall now outline some of the results of that paper. For definiteness we shall refer to amoebae responding to acrasin, but it will be clear that the discussion is valid for any organism which responds differentially to varying stimulus levels.

Suppose for simplicity that the organism moves either to the left or right in a step of length Δ. (The step may be identified with the extension of a pseudopod. Generalization to three-dimensional motion is straightforward.) Let receptors which are sensitive to the local acrasin level be located towards the ends of the organism, at a distance $\alpha\Delta$ apart. Assume that the average frequency ϕ of steps in a given direction depends only on the acrasin concentration ρ measured by the corresponding receptor. Then if $J(x)$ denotes the net number of cells per unit time which pass point x in the direction of increasing x we have

$$J(x) = \int_{x-\Delta}^{\Delta} \phi[\rho(s + \tfrac{1}{2}\alpha\Delta)]a(s)\,ds$$

$$- \int_{x}^{x+\Delta} \phi[\rho(s - \tfrac{1}{2}\alpha\Delta)]a(s)\,ds$$

where, as above, a is the cell density. Following common practice, we keep only the lowest order term in Δ. We thus obtain the approximation

(5.1)

$$J(x) \approx \Delta^2\{-\phi[\rho(x)]a'(x) + (\alpha - 1)\phi'[\rho(x)]a(x)\rho'(x)\}$$

where a prime denotes a derivative. Comparing with the one-dimensional version of (2.3) and (2.4),

(5.2) $$J = -\mu a' + a\chi\rho',$$

we find that

(5.3) $$\mu = \phi\Delta^2, \qquad \chi = (\alpha - 1)\phi'\Delta^2.$$

Note the striking relationship

(5.4) $$\chi(\rho) = (\alpha - 1)\mu'(\rho).$$

Of the observations which can now be made, we select the fact that the case $\alpha = 0$ corresponds to chemotaxis due to undirected effects on activity which arise from changes in chemical concentration sensed by a single receptor. This has been termed *chemokinesis* [7].

To proceed further, the step frequency $\phi(\rho)$ must be specified. In the absence of specific information we chose for illustrative purposes a simple threshold model involving $\xi(x)$ the instantaneous acrasin concentration at x. We assume that

$k =$ frequency of steps initiated at x when $\xi(x) > q$;

$k(1 - \bar{k}) =$ frequency of steps initiated at x when $\xi(x) < q$.

It is then easy to see that

$$\phi[\rho(x)] = \Delta^{-2}\mu[\rho(x)] = k\left\{1 - \bar{k}\int_{0}^{Q} F[\xi, \rho(x)]\,d\xi\right\}$$

where $F(\xi, \rho) d\xi$ is the probability that the acrasin level
is sensed to be between ξ and $\xi + d\xi$, given that the mean
acrasin concentration is ρ. Presumably the acrasin level
follows a Poisson distribution. If V denotes the effective
volume of the receptor then

$$\mu(\rho) = k\Delta^2 [1 - \bar{k} \sum_{N=0}^{N^*} (cV)^N e^{-cV}/N!]$$

where $N = \xi V$, and $N^* = QV$ is the threshold number
of molecules in the receptor. From $\chi = (\alpha - 1)\mu'$,
we find

$$\chi(\rho) \approx k\bar{k}\Delta^2(\alpha - 1) V(cV/N^*)^{N^*} e^{N^* - cV}(2\pi N^*)^{-1/2},$$

where we have used Stirling's formula.

As the acrasin concentration ρ increases from 0 to ∞,
μ increases monotonically from $k\Delta^2(1 - \bar{k})$ to $k\Delta^2$,
while χ increases from zero to a maximum

$$\chi_{\max} \approx k\bar{k}\Delta^2(\alpha - 1) V(2\pi N^*)^{-1/2}$$

and then decreases to zero again. The maximum value
of χ occurs when $\rho V = N^*$, i.e., when the average number
of molecules in the receptor equals the threshold number.
In our discussion of χ we have assumed that step size
is smaller than effective body length $(\alpha > 1)$ as is
appropriate for amoebae. Thus $\chi > 0$ and chemotaxis is
in the direction of increasing stimulus concentration as
observed.

Details and considerable further discussion can be
found in the paper of Keller and Segel [34]. Also to be
found there is a brief treatment of a simple model where
step length varies with concentration but overall step frequency remains constant. Again χ is found to be a
multiple of μ'. This illustrates the fact that quite different
microscopic models can give rise to the same phenomenological picture. On the macroscopic level, then, much

microscopic detail is irrelevant. There is of course great intrinsic interest in phenomena at the cellular, molecular, and (indeed) quantum mechanical levels, but I believe that each level is best treated mainly on its own terms. This hierarchical point of view is commonly held [35], [36, pp. 124–126], [37] but it does not seem fully appreciated in some biological circles.

Approximating random walk by diffusion is an idea that has frequently proved useful in biology. In addition to previously cited references, one might select for mention earlier work by Skellam [38] on population dispersal and a recent study by Blumenson [39] on the spread of cancer.

It has recently been brought to my attention that Patlak [40], [41] has studied the application of the random walk model to chemotaxis. By an analysis which is in several respects more general than that of Keller and Segel [34], Patlak arrives at an expression for the flux J which, like (5.2), is a linear combination of the cell density a and its derivative. The coefficients in this combination are complicated expressions involving terms which represent averages of the organism's speed, time between turns, distance between turns, persistence of direction, etc. The stimulus intensity (in our case chemical concentration) does not appear as a variable in Patlak's work, for he is interested in relating macroscopic flux with average values of microscopic path parameters. His success in doing this in various particular instances is only partial, because of the unavailability of the necessary detailed information. In his extensive discussion of various experiments, Patlak emphasizes that adaptation to a stimulus level is important in many contexts. The process is then non-Markovian and the random walk model becomes inappropriate. An approach by taking moments of a Boltzman-like equation would then be necessary.

In its emphasis on path parameters rather than stimulus levels Patlak's work [40], [41] supplements that of Keller and Segel [34]. Thus, if careful observations of an organism are made in conjunction with theoretical development, then it will be possible to deduce macroscopic flux expressions from details of microscopic motion. But I feel that direct measurement of phenomenological coefficients like χ and μ may provide a quicker route to understanding macroscopic phenomena.

6. Discussion

At the beginning of a *Scientific American* article on cellular slime mold, Bonner [42] remarked that some Soviet scientists visiting him at Princeton seemed to take only a polite interest in his work until he characterized it as an investigation of collective behavior in social amoebae. The field of research which we have been discussing can also be accurately and fashionably described as cellular ecology. Mathematicians seem particularly drawn to this field because of their propensity for revealing and clarifying order. It is appropriate to mention three distinguished mathematicians who have made contributions to the general area of the evolution of structure in development.

Morse [43] discussed Bonner's work on slime mold aggregation in an expository paper on equilibria in nature. He did not concern himself with mechanism but rather with a description of the aggregation pattern in terms of analogies with mountain tops and saddle points. Such a description gives theoretical restrictions on the possible patterns.

Thom [44] used the concept of structural stability in a general discussion of the emergence of spatio-temporal structure. His introduction of "elementary catastrophies" provides the foundation for an "art of models" which can provide conceptual guidance in a search for understanding of morphological processes.

Turing [45] studied the stability of homogeneous states of reacting and diffusing chemicals.[5] Although his models are fairly general in character, Turing presents a number of detailed results which are of considerable interest in connection with such matters as the formation of skin and tentacle patterns. It is a pity that he did not live to carry further his work in this area. He promised a machine-aided analysis of some nonlinear equations relevant to the problem of phyllotaxis (leaf formation).

To emulate Turing is a lofty goal for a mathematician who desires to make a contribution to the biological sciences. The work on Turing machines can be regarded as a fundamental attack on the problem of understanding the relation between brain function and structure, while the paper just described is a cornerstone in the theoretical foundations of developmental biology. The importance of Turing's work in logic probably leads many to regard him as a pure mathematician *par excellence*, but his 1952 work is a paradigm of applied mathematics. There mathematics is used as a means to the end of making a contribution to biology. This is attested to by the publication of Turing's work in Series B (Biological Sciences) of the Royal Society's *Philosophical Transactions*, rather than Series A (Mathematical and Physical Sciences). Realizing the importance of communicating his work to biologists, Turing included in his paper a section on required mathematical background. His remarks make it clear that he had made an effort to acquire a good understanding of the relevant biological facts. The paper is long, for there is considerable effort to make the arguments understandable. Turing's 1952 paper thus illustrates a four-fold way to successful biomathematics: (i) learn as much as you can of the bio-

---

[5] Turing's ideas have been refined and extended by various theoreticians [48]–[53] and have been examined from a biologist's viewpoint [54], [55].

logical background in some promising area, (ii) in consultation with experts, select a worthwhile problem in the solution of which mathematical concepts and techniques may prove helpful, (iii) use mathematics to obtain results of biological significance, (iv) take care to communicate your results so biologists can understand them.

Formulation and development of *general* conceptual models can lead to results of biological significance, but one must guard against the danger of excessive development of such models. If such development leads to interesting new mathematics, well and good. But if the mathematics is routine, the work is on all accounts inferior unless the biological consequences are interesting. Selecting for investigation particular biological problems is a way to avoid the error of valueless generalization, but one must of course attempt to select a particular problem which is representative, not parochial.

The present discussion has been centered on the aggregation of slime mold amoebae. One can never be sure of such things, but the discovery that the ubiquitous chemical cyclic $3'-5'$ AMP is an acrasin seems to lend great support to already prevalent views that slime mold aggregation is a particular manifestation of Turing's general ideas whose analysis can lead to a deeper understanding of development. The linear stability analyses which we have discussed seem to offer valid explanations for several aspects of aggregation but, as we have indicated, a number of extensions are required. Nevertheless, both the stability calculations and the analysis of travelling bands indicate that the continuum framework which we have employed can be of use in understanding collective cellular motion.

Acknowledgements

The author is grateful to the Army Research Office

(Durham) for partial support; to J. T. Bonner, E. F. Keller, and P. D. J. Weitzman for valuable comments on a draft version of this paper; to G. Gerisch for communicating his latest experimental results; and to Bonner, Gerisch, and B. Shaffer for permission to use figures.

REFERENCES

1. J. T. Bonner, *The cellular slime molds,* 2nd ed., Princeton Univ. Press, Princeton, N. J., 1967.

2. A. D. J. Robertson, *An analysis of developmental control in Dictyostelium discoideum,* Lectures on Math. in the Life Sciences, vol. 4, Amer. Math. Soc., Providence, R. I., 1971.

3. J. T. Bonner, D. S. Barkely, E. M. Hall, T. M. Konijn, J. W. Mason, G. O'Keefe III and P. B. Wolfe, *Acrasin, acrasinase, and the sensitivity to acrasin in Dictyostelium discoideum,* Developmental Biology **20** (1969), 72-87.

4. J. T. Bonner, *Aggregation and differentiation in the cellular slime molds,* Ann. Rev. Microbiol. (to appear).

5. J. Ashworth, *Cell development in the cellular slime mold Dictyostelium discoideum,* Symp. Soc. Exp. Biol., D. Davies (editor), Cambridge Univ. Press **25** (1971).

6. E. F. Keller and L. A. Segel, *Initiation of slime mold aggregation viewed as an instability,* J. Theoret. Biol. **26** (1970), 399-415.

7. G. S. Fraenkel and D. L. Gunn, *The orientation of animals,* Dover, New York, 1961.

8. F. Modabber, S. Morikawa and A. Coons, *Inflammation and Herpes simplex virus: release of a chemotaxis-generating factor from infected cells,* Science **170** (1970), 1104-1106.

9. B. M. Shaffer, *Integration in aggregating cellular slime moulds,* Quart. J. Microscop. Sci. **99** (1958), 103-121.

10. A. L. Steiner, D. M. Kipnis, R. Utiger and C. Parker, *Radioimmunoassay for the measurement of adenosine 3′, 5′-cyclic phosphate,* Proc. Nat. Acad. Sci. U. S. A. **64** (1969), 367-373.

11. T. M. Konijn, *Chemotaxis in the cellular slime molds. II. The effect of cell density,* Biol. Bull. **134** (1968), 298-304.

12. G. Gerisch, *Zellfunktionen und Zellfunktionswechsel in der Entwicklung von Dictyostelium discoideum. II,* Develop. Biol. **3** (1961), 685-724.

13. L. A. Segel, "Nonlinear hydrodynamic stability theory and its applications to thermal convection and curved flows," in *Nonequilibrium thermodynamics, variational techniques, and stability,* R. Donnelly, R. Herman and I. Prigogine (editors), Univ. of Chicago Press, Chicago, Ill., 1966, pp. 165-197.

14. _____, *Distant side-walls cause slow amplitude modulation of cellular convection*, J. Fluid Mech. **38** (1969), 203-224.

15. B. M. Shaffer, *The cells founding aggregation centres in the slime mould Polysphondylium violaceum*, J. Exptl. Biol. **38** (1961), 833-849.

16. G. Gerisch, *Die Bildung des Zellverbandes bei Dictyostelium minutum*, Roux' Archiv für Entwicklungsmechanik **155** (1964), 342-357.

17. _____, *Cell aggregation and differentiation in Dictyostelium*, Current Topics in Developmental Biology, A. A. Moscona and A. Monroy (editors), Academic Press, New York, **3** (1968), 157-197.

18. B. M. Shaffer, *Aspects of aggregation in cellular slime molds*: I. *Orientation and chemotaxis*, American Naturalist **91** (1957), 19-35.

19. M. H. Cohen and A. D. J. Robertson, *Acrasin, acrasinase, and periodic aggregative movements in polysphondylium*, J. Theoret. Biol. (to appear).

20. _____, *Chemotaxis and the early stages of aggregation in the cellular slime molds*, J. Theoret. Biol. **31**, (1971), 119–130.

21. _____, *Wave propagation in the early stages of aggregation of cellular slime molds*, J. Theoret. Biol. **31** (1971), 101–118.

22. T. M. Konijn, J. G. C. van de Meene, Y. Y. Chang, D. S. Barkely and J. T. Bonner, *Identification of adenosine-3'-5'-monophosphate as the bacterial attractant for myxamoebae of Dictyostelium discoideum*, J. Bacteriol. **99** (1969), 510-512.

23. J. R. Bowen, A. Acrivos and A. K. Oppenheim, *Singular perturbation refinement to quasi-steady state approximation in chemical kinetics*, Chem. Engrg. Sci. **18** (1963), 177-188.

24. F. G. Heineken, H. M. Tsuchiya and R. Aris, *On the Mathematical Status of the pseudosteady state hypothesis of biochemical kinetics*, Math. Biosci. **1** (1967), 95–113.

25. B. Edelstein, *Reaggregation of dissociated cells as a diffusion process*, J. Theoret. Biol. (to appear).

26. K. Bergman, P. V. Burke, E. Cerdá-Olmedo, C. N. David, M. Delbrück, K. W. Foster, E. W. Goodell, M. Heisenberg, G. Meissner, M. Zalokar, D. S. Dennison and W. Shropshire, Jr., *Phycomycetes*, Bacteriol. Rev. **33** (1969), 99-157.

27. J. Adler, *Chemotaxis in bacteria*, Science **153** (1966), 708-716.

28. _____, *Chemoreceptors in bacteria*, Science **166** (1969), 1588-1597.

29. E. F. Keller and L. A. Segel, *Travelling bands of chemotactic bacteria: a theoretical analysis*, J. Theoret. Biol. **30** (1971), 235-248.

30. J. Adler and M. Dahl, *A method for measuring the motility of bacteria and for comparing random and non-random motility*, J. Gen. Microbiol. **46** (1967), 161-173.

31. J. L. Smith and R. N. Doetsch, *Studies on negative chemotaxis and the survival value of motility in Pseudomonas fluorescens*, J. Gen. Microbiol. **55** (1969), 379-391.

32. I. M. Gel'fand, *Some problems in the theory of quasilinear equations,* Uspehi Mat. Nauk **14** (1959), no. 2 (86), 87-158; English transl., Amer. Math. Soc. Transl. (2) **29** (1963), 295-381. MR **22** # 1736; MR **27** # 3921.

33. A. Kolmogoroff, I. Petrovsky and N. Piscounoff, *Étude de l'équation de la diffusion avec croissance de la quantité de matière et son application à un probleme biologique,* Bulletin Mathématique (Moscow Universitet), Séries Internationale, Section A. **1** (1937), 1-26.

34. E. F. Keller and L. A. Segel, *Model for chemotaxis,* J. Theoret. Biol. **30** (1971), 225-234.

35. C. Grobstein, *Levels and ontogeny,* Amer. Sci. **50** (1962), 46-58.

36. R. Feynman, *The character of physical law,* MIT Press, Cambridge, Mass., 1967.

37. P. Weiss, "The living system: determinism stratified," in *Beyond reductionism–new perspectives in the life sciences,* The Alpbach Symposium 1968, A. Koestler and J. R. Smythies (editors), Macmillan, New York; reprinted in Studium Generale **22** (1969), 361-400.

38. J. G. Skellam, *Random dispersal in theoretical populations,* Biometrika **38** (1951), 196-218. MR **13**, 263.

39. L. E. Blumenson, *Random walk and the spread of cancer,* J. Theoret. Biol. **27** (1970), 273-290.

40. C. S. Patlak, *Random walk with persistence and external bias,* Bull. Math. Biophys. **15** (1953), 311-338. MR **18**, 424.

41. _____, *A mathematical contribution to the study of orientation of organisms,* Bull. Math. Biophys. **15** (1953), 431-476.

42. J. T. Bonner, *Differentiation of social amoebae,* Scientific American **201** (1959), 152-162.

43. M. Morse, *Equilibria in nature – stable and unstable,* Proc. Amer. Philos. Soc. **93** (1949), 222-225. MR **10**, 586.

44. R. Thom, *Topological models in biology,* Topology **8** (1969), 313-335. MR **39** # 6629.

45. A. M. Turing, *The chemical basis of morphogenesis,* Philos. Trans. Roy. Soc. London Ser. B **237** (1952), 37-72.

46. J. T. Bonner, *Hormones in social amoebae and mammals,* Scientific American **220** (1969), 78-91.

47. H. Rasmussen, *Cell communication, calcium ion, and cyclic adenosine monophosphate,* Science **170** (1970), 404-412.

48. J. I. Gmitro and L. E. Scriven, *A physicochemical basis for pattern and rhythm,* Intracellular Transport, K. B. Warren (editor), Academic Press, New York. 1966.

49. I. Prigogine and G. Nicolis, *On symmetry-breaking instabilities in dissipative systems,* J. Chem. Phys. **46** (1967), 3542-3550.

50. I. Prigogine, *Dissipative structures in biological systems,* Second Internat. Conference on Theoretical Physics and Biology, Institut de la Vie, Versailles, 1969.

51. H. G. Othmer and L. E. Scriven, *Interactions of reaction and diffusion in open systems,* Indust. Engrg. Chem. Fund. **8** (1969), 302-313.

52. B. B. Edelstein, *Instabilities associated with dissipative structure,* J. Theoret. Biol. **26** (1970), 227-241.

53. M. H. Cohen, *Models for the control of development,* Proc. Twenty-Fifth Sympos. Control in Growth and Development (1971), Society of Experimental Biology and British Society of Developmental Biology (to appear).

54. J. M. Smith, *Continuous, quantized and modal variation,* Proc. Roy. Soc. B **152** (1960), 397-409.

55. J. M. Smith and K. C. Sondhi, *The arrangement of bristles in Drosophila,* J. Embryol. and Exp. Morph. **9** (1961), 661-672.

56. B. M. Shaffer, *The Acrasina,* Advan. Morphogenesis **2** (1962), 109-182.

57. J. T. Bonner, *Physiology of development in cellular slime molds* (*Acrasiales*), Encyclopedia of Plant Physiology, W. Ruhland (editor), vol. XV/1, Springer-Verlag, Berlin-Heidelberg-New York, 1965, pp. 612-640.

58. J. T. Stuart, *On the cellular patterns in thermal convection,* J. Fluid Mech. **18** (1964), 481-498.

QUANTITATIVE ANALYSIS OF THE DEVELOPMENT OF CELLULAR SLIME MOLDS*

By

ANTHONY ROBERTSON

University of Chicago

*Supported in part by the Alfred P. Sloan Foundation and the Otho S. A. Sprague Memorial Institute.

Contents

1. Introduction

In this paper I want to show how the application of a minimal amount of quantitative reasoning has helped us to understand the control of some of the developmental processes in the cellular slime molds.

In biology the most interesting advances are usually made as a result of the propitious choice of an experimental system, and of the level at which to model its behavior. For example, Hodgkin and Huxley were able to develop a theory of the action potential because they had a suitably large and robust preparation, the squid giant axon [1]. Similarly, much progress in molecular biology depended on an understanding of the genetics of bacteria and their viruses [2]. Development itself is so complex, and its study has been so intense, that it is impossible to assimilate the known facts, let alone interpret them. I feel, therefore, that two steps are needed if our understanding of the control of development is to advance at all rapidly. Firstly, a logical structure for the subject, to which known facts can be related, is essential; secondly, we must choose for study systems that are simple enough to understand, but sufficiently complex to be interesting. Given these prerequisites we must also make a clear decision about the level of organization we wish to investigate. I intend, in this paper, to consider events only at or above the cellular level. It is my prejudice that this is appropriate for the study of development now and that it is premature to attempt an analysis of developmental molecular biology.

At the cellular level there are six developmental processes: growth, division, movement, differentiation, contact formation and secretion [3]. These are precisely controlled in space and time so that adult organisms of the same genotype are closely similar in phenotype. This reliable control implies the existence of a developmental control system. The classification and possible nature of such systems has recently been fully discussed [4].

In the next section of this paper I will describe a model for the control of development, which is used to illuminate what we know of the life cycle of some of the cellular slime molds, described in §3. In §4, I show how quantitative analysis has led to the planning and interpretation of recent experiments on the cellular slime molds. Finally, in §5 I discuss the relationship of development in the cellular slime molds to development in the Metazoa.

2. Models for the control of development

A model for developmental control must contain a means of supplying both positional and temporal information to cells in an embryo [5]. That is, it must provide both a map and a clock [6]. Information from the map and the clock can then be interpreted by cells which proceed with those developmental processes appropriate to the developmental time and the cells' position within the embryo.

The classical embryological model for developmental control depends on a gradient [7] (Figure 1). In the simplest case cells can detect their position in a concentration gradient of a substance, a morphogen, and act accordingly. This model has recently been extended to show that there is indeed time to set up a morphogen gradient by diffusion from a set of source (manufacturing) cells towards a set of sink (destroying) cells on the usual developmental distance scale [8]. However, it is hard to make a gradient which regulates, that is, which is size-independent. As the production of size-independent structure is a fundamental property of embryos the

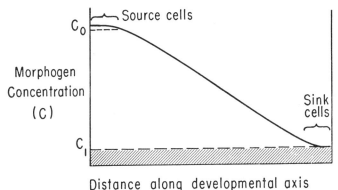

FIGURE 1. Linear concentration gradient of morphogen produced by cells in the source region and destroyed by cells in the sink region. When $c < c_0$ locally source cells manufacture morphogen, when $c > c_0$ locally sink cells destroy morphogen. The gradient is maintained by diffusion.

control systems underlying their development must also regulate. Furthermore, it is quite difficult to produce spatially periodic patterns by diffusion without making somewhat arbitrary assumptions about the kinetics of the system (see Turing, 1952) [9].

A radically different kind of model (Figure 2), proposed by Goodwin and Cohen [6], obviates these two difficulties, although it is complex, involving three periodic events. Goodwin and Cohen suggest that cells in an embryo are autonomously periodic when isolated from their neighbors, some process within them oscillating with a period of minutes. They call this process the S-event; it may be complex, but it contains a brief component, the signalling aspect, which can signal from cell to cell via the tight junctions known to couple cells within early embryos [10]. The signalling aspect can advance the phase of the S-event in a neighboring cell, allowing entrainment between coupled cells. In order to get stable entrainment there must be a frequency gradient falling from a dominant, or pacemaker, region, and each cell must, after being signalled, remain refractory for a significant portion of the period, T, of the S-event, otherwise it would be restimulated by reflections of its own signal. Given these constraints a wave of the S-event, the S-wave, will propagate through a "field" of coupled cells from the pacemaker, or "organiser" region. The S-wave will be repeated periodically. Its velocity will depend on the delay involved in intercellular signalling, Δt_S. Coarse positional information is already present in the frequency gradient, which is supposed to derive from an initial gradient, e.g. of yolk distribution, in the fertilized egg. Fine positional information can be provided by the phase difference $\Delta \phi_{PS}$ between the occurrence of the S-event at a point in the embryo and a second, P-event, which propagates from the pacemaker region with a different velocity. The system automatically produces periodic positional information when the phase difference $\Delta \phi_{PS}$ exceeds 2π plus the phase difference between P and S at the origin. Furthermore, the system can be made to regulate. These features are fully discussed in Goodwin and Cohen's paper [6]. I do not wish to imply that their model is "correct" or indeed necessarily more biologically realistic than a gradient model, but I do want to emphasize that they clearly proved that such a system would provide the positional and temporal information necessary for the control of development.

In a simpler version of their model, the single-event model, positional information is supplied by the difference in phase between a propagating signal and the local phase of the same process [11]. Local phase is reset by the signal, but as there is an underlying frequency gradient positional information can be supplied by the difference between actual phase and what would be local phase if the process were allowed to proceed autonomously.

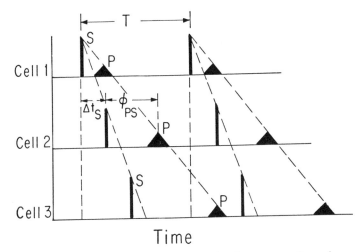

FIGURE 2. Goodwin-Cohen model showing propagation of signalling aspects of S- and P-events along a line of three cells. Note monotonic increase of ϕ_{PS}, the phase difference between S- and P-events. T is the period of the pacemaker S-event. Δt_S is the delay in transmission of S from cell to cell.

Thus, global time can be compared with local time.

Clearly, in searching for real developmental control systems both "quasistatic" (gradient) and periodic models must be considered.

We started to investigate development of the cellular slime molds because we knew that one of the developmental processes, movement, was periodic in time in one species, Dictyostelium discoideum [13]. Therefore, we were in fact stimulated by the features of a novel model, although it soon became apparent that it was more important to find out what actually happens in slime mold development than to try to fit the known facts to a particular model!

I shall now describe the life histories of typical cellular slime molds and show how we have been able to understand a little of the developmental control processes involved.

3. The life cycles of the cellular slime molds

All cellular slime mold amoebae feed on bacteria [12], [13], [14] (see Figure 3). When food is available they grow and divide; when it is consumed they aggregate. The amoebae live in a thin ($\sim 1\mu$) aqueous film on the surface of their substrate, which might be soil, forest leaves or dung. They find their food by chemotaxis, moving up a gradient of attractant released by the food bacteria. In the absence of an attractant movement is random, or diffusive; in its presence movement is directed. The organelles involved in movement, pseudopods, can apparently be produced from any part of the membrane surface. When the supply of bacteria is exhausted the amoebae first move rapidly at random, then become quiescent and more or less stationary. In some species, of which the best known is Dictyostelium minutum, a few single amoebae, the founder cells, which are round and remain stationary, begin to secrete an attractant [15], probably cyclic AMP. This leads to the production of gradients with founder cells at points of highest concentration. Neighboring amoebae move towards the founder cells. In time-lapse films one can see a sharp transition between random movement and directed movement towards a founder cell [16]. The transition occurs on a circle of increasing diameter (up to $\sim 300\mu$) centered on a founder cell. Those amoebae which aggregate and come in contact with the founder are themselves transformed to secreting cells, after a delay of about 1/2 hour. Thus aggregates with circular peripheries grow until all the amoebae in the vicinity have been attracted. Some of

FIGURE 3a (facing page). Life cycle (with Figures 3b, c and d) of D. discoideum (photographed and prepared by David Drage). A – F, spore germination; G, vegetative amoeba containing partially digested bacteria and showing nucleus and contractile vacuole. Length marks: 10μ.

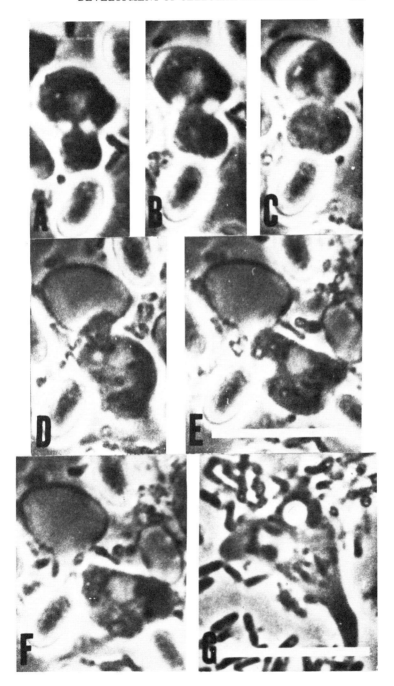

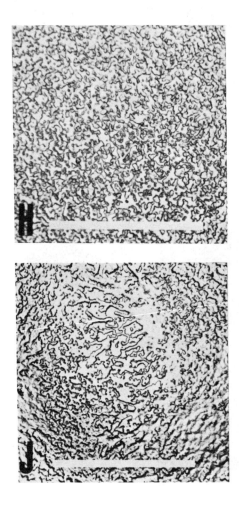

FIGURE 3b. H, field of amoebae in early interphase one hour after centrifugation; J, six hours later showing beginning of aggregation. Length marks: 1 mm.

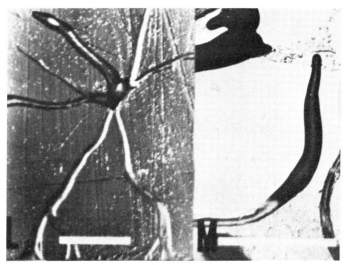

FIGURE 3c. K, two hours later than J, showing aggregation with tip and streams; L, later aggregate with slug leaving, six hours later than K; M, migrating slug with slime sheath collapsing behind. Length marks: K, 1mm; L and M, 1/2mm.

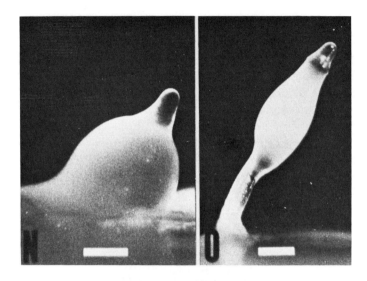

FIGURE 3d. N, early culmination about one hour from the end of migration; 0, fruiting body erecting about 3 hours after N; P, eight hours after 0, fruiting body with stalk, spherical spore mass and tip differentiating into spores.
Length marks: N and 0, 1/10mm; P, 1/2mm.

the neighboring amoebae are themselves transformed into founder cells without beginning to move. (It is possible that movement and secretion are mutually exclusive functions of the cell membrane and that a cell capable of producing pseudopods from any point on its membrane surface cannot secrete attractant until it has become stationary and it has reorganized, losing the capacity for further movement.) At the end of aggregation some of the cells differentiate to take part in stalk formation. The remaining amoebae differentiate into spores and are held aloft on the stalks.

D. minutum is typical of those species of cellular slime mold in which movement, a developmental process, is controlled by a concentration gradient of an attractant. The gradient is initially produced by a single founder cell; further amoebae are transformed into secreting cells only when they have reached an aggregate, made contact with previously arrived cells, and remained stationary for a time. The resulting simple morphology with a circular aggregate and simple fruiting bodies produced at the site of aggregation is apparently typical of this group.

In D. discoideum both the signal controlling aggregation and the resulting morphogenetic movement are more complex. When feeding, and therefore growth and most cell division, cease, the amoebae first move rapidly at random for about half an hour, and then, apparently because bacterial attractants are no longer present in the aqueous film, become stationary for a period. This period, interphase, lasts for 6 to 8 hours [15]. During interphase the amoebae differentiate into polar cells [17], [18]. In the vegetative (feeding) phase each amoeba could move by extending pseudopods from any portion of its membrane surface, but at the end of interphase only one end of the amoeba can produce pseudopods.

We believe also that the membrane at this end contains the receptors for cyclic AMP and that cyclic AMP can only be released from the other, posterior, end [18]. At the end of interphase an amoeba is capable of releasing pulses of cyclic AMP as well as responding to such a pulse by both moving towards the signal source and producing a cyclic AMP pulse. These abilities develop at different relative times during interphase, and at different rates for particular amoebae. From our observations of films taken during interphase, as well as from the preliminary results of experiments with an artificial signal source, we have concluded that the ability to move in response to a cyclic AMP signal develops before the ability to produce such a signal.

At the end of interphase some amoebae, randomly distributed in the population, begin to produce pulses of cyclic AMP. Presumably these are the amoebae that have differentiated most rapidly. The cyclic AMP diffuses away from these cells. The pulse's amplitude is reduced both by diffusion and by the action of an extracellular phosphodiesterase, which converts the cyclic AMP to linear 5′ AMP, which is not effective as a chemotactic agent. When a neighboring amoeba is stimulated by a supra-threshold concentration of cyclic AMP it responds in quite a complex way (Figure 4). After a delay it produces its own cyclic AMP pulse, which relays the signal centrifugally, then it begins a centripetal movement step [19]. As the original signalling cell repeats its pulse periodically the amoebae aggregate. Where population density is locally high amoebae go towards their nearest neighbors, rather than directly towards the center. Streams therefore form [14]. We believe these to be characteristic of those species in which each amoeba acts as a local source for the propagation of an aggregative signal in contrast to those that migrate up a con-

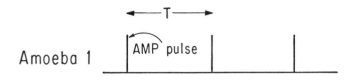

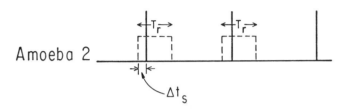

FIGURE 4. Time course of D. discoideum aggregative signal.
T = period of cyclic AMP pulse ($\cong$ 300 sec.); Δt_S = intracellular
delay ($\cong$ 15 sec.); Tr = refractory period ($\cong$ 100 sec.). The
cyclic AMP pulse is short, duration $\leq$ 2 sec.

centration gradient produced by a single source. During
aggregation signs of further differentiation appear. Some
cells, at the top of what is a growing hemispherical mass,
form a tip with distinctive histochemical and physiological
properties [20]. Cells in the tip, and possibly others,
begin to secrete a muco-polysaccharide slime that flows
down as a liquid, hardening to form a cone inverted over
the aggregation center. As amoebae continue to enter
the aggregation via the streams, and as the slime is only
liquid near the tip, the resulting pressure can only be
released by the erection of a column of cells, bearing
up the secreting tip. The column therefore becomes
covered with hardened slime. When sufficiently tall it
becomes unstable, falls over to form the slug stage, and
crawls off leaving the collapsed slime sheath behind.

Many of the features of the signal (Figure 4) and the
response to it may be revealed by a study of early aggre-

gation. In particular, aggregation centers are not stable at first, but, as amoebae with higher autonomous frequencies emerge from interphase, these entrain their neighbors and become new centers. Also boundaries form between the territories controlled by neighboring centers. Signals do not propagate through these boundaries, which are refractory boundaries as discussed in the next section [15]. Finally the polarity of individual amoebae is very clearly demonstrated, particularly by the way in which they enter the streams [21].

The slug moves towards light and up a humidity gradient [13]. It is not distracted by passing through fields of bacteria [13]. There is clear evidence, beautifully collected and summarized by Takeuchi [20], and recently confirmed by Bonner, et al. [22], that by the time the slug has been formed "sorting-out" has occurred. This term refers to the arrangement of cells within the slug, implying that those cells, in the front third of the slug, that ultimately contribute to the stalk of the fruiting body, are already in place, as are those cells which will become spores and which are by now located in the rear two-thirds of the slug. This has been demonstrated in several ways. Firstly, by the time the slug has formed the prespore and prestalk regions show specific histochemical reactions [20] and specific ultrastructure [23]. Secondly, those cell properties that have been measured, e.g. density, show a random distribution of values amongst amoebae in a pre-aggregation field, but a graded distribution within the slug [20], [22]. Final differentiation, however, has not occurred; we postulate that this differentiation depends on an individual cell's interpretation of its position within a gradient in the slug. An obvious candidate for the graded property is autonomous pulse frequency, which one would, of course, expect to be correlated with other properties related to metabolic rate. I shall discuss this briefly in the next section.

After migrating for a time the slug stops, possibly because it can no longer make the slime within which it moves, possibly because the slime becomes too hard, or possibly because the amoebae within the rear of the slug, which were initially joined in strings, lose some of their contacts as they differentiate into spores and can no longer support the movement of the slug. At this stage, all the cells collect to form a squashed spherical mass under the tip. Some of the prestalk cells in the center of the mass just below the tip elongate and become vacuolated, forming the beginning of the stalk [13]. This extends down through the mass until it makes contact with the base-plate, a small group of cells initially at the rear of the slug. When this happens further movement of prestalk cells to the axis of the fruiting body, and their elongation to form the next segment of the stalk, result in elevation of the prespore mass above the substrate. Finally, when the supply of prestalk cells is exhausted a fruiting body consisting of a spherical mass of spores, perhaps 10^5 in number, supported on a stalk up to 2 mm. long, has been formed. The spore head dries out and the polysaccharide-coated spores can be blown away. If they reach a suitable substrate they will germinate under specific conditions of humidity and chemical composition of the environment, completing the life cycle [24].

4. Quantitative experimental and theoretical analysis of the life cycle

In this analysis I shall consider only those phases of the life cycle that begin after the cessation of feeding, with the exception of the mechanism of movement by vegetative amoebae. Vegetative amoebae of both the type-species, D. minutum and D. discoideum, move at random in the absence of an external signal. That is, they have a diffusion constant D that can be measured by measuring the increase with time of an area circumscribed by the line connecting the most peripheral members of an initially closely neighboring set of amoebae. Although we have not yet done this quantitatively the change in D can be clearly seen as a chemotactic source becomes effective. Observations of time-lapse films show that the movements of D. minutum amoebae towards either a food source or a founder cell are quite directed [15]. Therefore presence of a chemotactic agent reduces D markedly. A model which has this property has been developed by Cohen [3].

Each vegetative movement normally consists of the following events. A filopod, a membrane-covered cytoplasmic extension about $1/2\mu$ in diameter and about $5\text{-}10\mu$ long is projected normal to the membrane surface (r.m.s. deviation $= 6°$). This may wave about and then be retracted, or it may anchor to the substrate. In the latter event a pseudopod forms by the ballooning out of the membrane at the filopod base towards the filopod tip, with cytoplasmic flow around the internal rigid structure of the filopod. In other amoebae filopods have been shown to contain rigid skeletal elements, possibly projected from the cytoplasmic skeleton of the cell by the polymerization in situ of protein monomers to form micro-filaments, but we have so far found only sparse evidence for filopodial microfilaments in the cellular slime molds.

After pseudopod formation, contraction of the rear of the cell results in further cytoplasmic flow into the pseudopod and leads to definite displacement of the center of gravity of the cell. Eventually the amoeba moves over the initial anchorage point of the filopod which may become stretched out behind the forward-moving cell and may even be broken off. We have observed that in a randomly-moving amoeba the sites of successive pseudopod formation are uncorrelated in space and time, except that the membrane can only support 5 or 6 pseudopods simultaneously. The minimum spacing between pseudopod axes at the cell margin is therefore about 5μ for a cell $8\text{-}10\mu$ in diameter. Filopods can be much closer together with a minimum separation of a micron or less. Therefore local production of a filopod must locally inhibit the initiation of further filopods which might lead to pseudopod formation.

In a chemotactically directed amoeba, however, one part of the membrane surface becomes dominant in pseudopod production, leading, because filopods are initiated at right angles to the membrane surface, to movement in a straight line up the gradient of chemotactic agent. In most theories of chemotaxis the chemotactic agent increases the frequency of some self-excited membrane process which leads to pseudopod production. Clearly this would lead to an increase in D and therefore net movement down the chemotactic gradient. We can get round this difficulty by assuming that the chemotactic agent excited filopod formation at that point where the membrane surface first tangent to the contour of constant concentration equal to the threshold concentration for chemotaxis, and that there is an inhibition of pseudopod formation elsewhere, inhibition spreading away from the initial pseudopod and the inhibitory time-constant being slightly longer than the time constant for the movement step. The first excited site will then be the first to recover for subsequent stimulation by a supra-threshold concentration of chemotactic agent. Therefore one site of pseudopod production, that is one portion of the membrane surface, can become entrained in the presence of a chemotactic agent, resulting in a decrease in D and in straight-line movement up a concentration gradient. We believe that this model also suffices to account for the early stages of aggregation of D. minutum.

Although the vegetative movements of D. discoideum appear similar to those of D. minutum, the aggregative movements are much more complex, as is their control. The post-interphase amoebae are polar, with a motile and receptor front end, and a signalling rear end [18]. They produce brief pulses of cyclic AMP [19]. The critical density of amoebae below which aggregation cannot occur is at an interamoeba spacing of 60-70μ. A molecule the size of cyclic AMP would diffuse over this distance in about 2 seconds which is very brief compared both with the period of the signal, about 300 seconds, and with the velocity of propagation of the wave of inward movement described in §3 [19]. Because of this discrepancy in velocities we were forced to assume that the rate-limiting step in signal propagation is an intracellular delay of about 15 seconds between being signalled and signalling. Interestingly enough this delay is also consistent with the velocity of signal propagation in later stages of the life cycle when the amoebae are joined by antero-posterior junctions. To ensure unidirectional centripetal propagation of the wave of signalling and movement in response to the signal we have had to assume a refractory period beginning with receipt of the signal and continuing for about 100 seconds, roughly the duration of the intracellular delay plus the movement step initiated by the signal. Because the signal is very brief, being reduced by both diffusion and the extra-cellular phosphodiesterase, the movement provoked by such a signal is an all-or-nothing event [18], [19]. We have also been able to calculate that so much cyclic AMP is used that the amoebae must take up the 5′ AMP produced and recycle it; otherwise they would run out of the signalling agent [19]. This prediction could clearly be tested by following the history of the labelled nucleotide. Other details of the mechanisms of wave propagation and chemotaxis in the early stages of aggregation of D. discoideum may be found in our papers. The picture includes the ideas of a pulsatile signal propagated through the field at roughly constant velocity, refractoriness and all-or-nothing response to the signal, and entrainment by groups of "pacemaker" amoebae.

Our initial consideration of wave propagation in early aggregation led us to a value for $\mathcal{N}$, the number of molecules of cyclic AMP released in a pulse by a single amoeba, of about 3×10^9 [19]. Using this number we have been able to simulate the action of an aggregative center by the periodic electrophoretic release of pulses of cyclic AMP from a microelectrode. This appears to mimic the action of a real center exactly and should provide a powerful tool for the elucidation of developmental control in D. discoideum as well as in more complex embryos where there is also some reason to believe that transmitters are involved in developmental control.

The signalling mechanisms used by the cellular slime molds are certainly reflected in their morphologies during and after aggregation. We have begun to investigate this quantitatively. The major result is that streaming is always associated with a system in which each amoeba acts as a propagator of the signal, whether it is pulsatile or continuously produced. Where only the center and those amoebae which come into contact with the center produce the signal, no streams form. The role of phosphodiesterase is also interesting [25], [26]. It is produced (Bonner, personal communication) by those amoebae which use a pulsatile signal, but it is apparently not produced by the one smooth aggregator tested, D. minutum. This is suggestive, although there is not yet enough information in support of an absolute correlation. Presumably in a pulsatile system the phosphodiesterase is needed to remove the cyclic AMP in order to prevent the saturation of its receptors, which would lead to loss of directionality, and to allow it to be recycled.

On a microscopic scale we have observed D. discoideum amoebae as they approach a stream. They make very specific movement towards, and contact with, the rear portion of an amoeba in the stream. These contacts involve quite small membrane areas, with a diameter of perhaps $1/2\mu$, and there may be several regions of contact between the front and back membrane portions of two cells. The amoebae therefore tend to join together in strings. Similar strings of amoebae are found in the slug. The contacts forming the strings are EDTA resistant [15], [27], but those holding the strings together laterally are not. In the slug the latter appear as membrane interdigitations of an amoeba with its lateral neighbors [28].

The movements after aggregation have not yet been analysed fully. However, all those we have observed in D. discoideum have a periodic component with a modal period of about five minutes. Furthermore, all such movements appear to be controlled by a well-defined tip. If the tip is removed at any stage, coordinated movement ceases until a new tip has differentiated; the tip also seems to be necessary for regulation to proceed in isolated portions of transected slugs.

We have therefore started a series of experiments, as illustrated in the matrix of Figure 5. In all those entries marked +, we have performed the experiment of transplanting a tip of given developmental age into the given developmental stage of D. discoideum. In all cases so far tried the tip takes over a portion of the field into which it is transplanted. We have not done enough experiments to supply statistics of success and failure,

DEVELOPMENTAL AGE OF TIP

	Early to Late Aggregate	Conus	Slug	Fruiting Body
Vegetative amoebae				
Interphase amoebae (2 hr. intervals)				
2 hr.				
4 hr.				
6 hr.			+	+
8 hr.				
Post-Interphase amoebae	+		+	+
Aggregate	+		+	
Conus				
Slug			+	
Fruiting Body			+	+

FIGURE 5. Matrix showing projected tip transplantation experiments. An entry of + indicates an experiment already performed.

or to allow us to make quantitative statements about the signal propagation from the tips, but in all cases so far it appears qualitatively similar to signal propagation from an ordinary center. That is, the signal attracts competent, post-interphase amoebae with a period of approximately five minutes. We know that this periodicity is a property of the signal and not of the response system both from experiments by Gerisch who placed centers of a D. discoideum mutant which secretes cyclic AMP continuously into fields of competent wild-type amoebae [15] and from our own preliminary results with the artificial center.

We have begun quantitative analyses at later stages in the life cycle, in particular of slug movement and of fruiting body erection, but we do not yet have quantitative theories of these processes. Both, however, are consistent with the model of a pacemaker tip controlling movement by the periodic production of a signal which propagates with velocity limited by the intracellular delay of 15 seconds and which is similar to the early aggregative signal.

A problem we have so far avoided is that of differentiation and its regulation, except to observe that the presence of an active tip is probably a prerequisite for both. Removal of the tip of the slug delays both processes, and isolated slug portions do not begin to regulate for about one hour [29], by which time a new tip has formed.

5. Discussion

I want to make four points about the material presented in this paper.

1. It is important to choose a simple organism for the study of development. In particular each developmental process should be manifested in a simple fashion, with good spatial and temporal separation from the manifestation of the other processes. These conditions are satisfied by the choice of the cellular slime molds.

2. The construction of simple quantitative theories about processes at each stage of development gives immediate insight into the nature of the control system

involved. For example, it allows a much better understanding of the signalling system and it makes clear the
reasons for some morphological variations. It also suggests
phenomena that should be looked at and that may be
diagnostic for the identification of the appropriate model
for the control of development in the organism concerned.
For example, one should clearly look for periodic phenomena in the development of D. discoideum as these
may be related to developmental control at all stages
of the life cycle.

3. The processes of development in the cellular slime
molds are the same as those in complex Metazoa. Therefore the cellular slime molds are good model organisms;
and, therefore, we should now look for simple situations
in more complex embryos where similar control mechanisms might exist and where their action could most
easily be discerned. For example, one might attempt
to control the migration of the mesoderm cells that
form the heart rudiments in the chick with a microelectrode. This migration has a periodic component
$(T \cong 10$ minutes) [30] and seems qualitatively similar to
that of D. discoideum amoebae. Many other systems,
where transmitter action is already suspected, suggest
themselves. Some are listed in our "periodic table" [11].

4. Finally, there are strong analogies between the
control system of D. discoideum and the Metazoan
nervous system. A diffusible transmitter propagates a
rectified message and is removed by an enzyme; intracellular propagation is at constant velocity; the signal
is pulsatile; and the "field" at each developmental stage
acts as an excitable medium. The analogies are strong
enough to reinforce the postulate that nervous systems,
and other systems for the control of organ function,
may evolve both in phylogeny and ontogeny, as speciali-
zations of developmental control systems [6].

Acknowledgements

The work I have discussed has all been done in collaboration with Morrel Cohen to whom I owe much for his inspiration and insight. David Drage performed many of the experiments mentioned and Diane Wonio assisted in many aspects of the experimental work. I am also grateful for the hospitality of Professor E. R. Caianiello (C. N. R. Arco Felice, Naples) and Professor L. Wolpert (London).

REFERENCES

1. A. L. Hodgkin, *The conduction of the nervous impulse,* Liverpool Univ. Press, Liverpool, 1965.

2. J. D. Watson, *Molecular biology of the gene,* 2nd ed., Benjamin, New York, 1970.

3. M. H. Cohen and A. Robertson, *Proceedings of the I. U. P. A. P. conference on statistical mechanics* (Chicago, 1971), S. Rice, K. Freed and J. Light, Eds., Univ. of Chicago Press, Chicago, Ill. (to appear).

4. M. H. Cohen, Symp. Socs. Exp. and Dev. Biol. (to appear).

5. L. Wolpert, J. Theoret. Biol. **25** (1969), 1.

6. B. C. Goodwin and M. H. Cohen, J. Theoret. Biol. **25** (1969), 49.

7. C. M. Child, *Patterns and problems of development,* Univ. of Chicago Press, Chicago, Ill., 1941.

8. F. H. C. Crick, Nature **225** (1970), 420.

9. A. M. Turing, Phil. Trans. Roy. Soc. B **237** (1951), 37.

10. E. J. Furshpan and D. D. Potter, Current Topics Dev. Biol. **3** (1968), 95.

11. M. H. Cohen, *Lectures on mathematics in the life sciences,* vol. 3, American Mathematical Society, Providence, R. I., 1971.

12. J. T. Bonner, *The cellular slime molds,* 2nd ed., Princeton Univ. Press, Princeton, N. J., 1967.

13. K. B. Raper, J. Elisha Mitchell Sci. Soc. **56** (1940), 241.

14. B. M. Shaffer, Adv. Morphogen. **2** (1962), 109.

15. G. Gerisch, Current Topics Dev. Biol. **3** (1968), 157.

16. J. T. Bonner, Unpublished film kindly lent to the author.

17. _____, Symp. Soc. Exptl. Biol. **17** (1963), 341.

18. M. H. Cohen and A. Robertson, J. Theoret. Biol. **31** (1971), 119.

19. _____, J. Theoret. Biol. **31** (1971), 101.

20. I. Takeuchi, in "Nucleic acid metabolism, cell differentiation and cancer growth," Edited by E. V. Cowdrey and S. Seno, Pergamon Press, New York, 1969.

21. B. M. Shaffer, in "Primitive motile systems in cell biology," Edited by R. D. Allen and N. Kamiya, Academic Press, New York, 1964.

22. J. T. Bonner, E. Sieja and E. M. Hall, J. Embryol. Exp. Morph. 25 (1971), 457.

23. H. R. Hohl and S. T. Hamamoto, J. Ultrastructure Res. 26 (1969), 442.

24. D. A. Cotter and K. B. Raper, Proc. Nat. Acad. Sci. U. S. A. 56 (1966), 880.

25. B. M. Shaffer, Nature 177 (1953), 975.

26. Y.-Y. Chang, Science 160 (1968), 57.

27. R. L. De Haan, J. Embryol. Exptl. Morphol. 7 (1959), 335.

28. Unpublished observations of electron-micrographs taken by D. Drage and R. Lymn.

29. Unpublished observations of our films by M. H. Cohen and A. Robertson.

30. R. L. De Haan, Unpublished film of development in the chick.

AN ANALYSIS OF THE RETINO-TECTAL PROJECTION
OF THE AMPHIBIAN VISUAL SYSTEM

By

B. C. GOODWIN

University of Sussex, England

Introduction

This paper contains an analysis of a body of experimental data relating to the embryological processes leading to the establishment of a neighbourhood-preserving mapping of the retina onto the appropriate part of the brain (the optic tectum) in the South African clawed toad, *Xenopus laevis*. I am concerned with comparisons between the mappings predicted by two basically different types of model, to see how far one can go at the present state of experimental method in choosing between them. A complete description of the experimental techniques used and the properties of this projection (used in this context as equivalent to mapping), together with a great wealth of further information about the pattern of neural interaction in embryonic systems, is given in a recent book by R. M. Gaze [3]. The experimental material analyzed in this paper comes from work carried out in Dr. Gaze's laboratory, and I would like to acknowledge my indebtedness to him and his colleagues for their generosity in providing unpublished material and for very useful discussions about the problem.

The retino-tectal projection

Put very simply, the particular problem I want to consider is how the 50,000 or so nerve fibres which grow from the retinal neurones of the eye in *Xenopus* make ordered connections with about the same number of neurones in the optic tectum. In *Xenopus* and amphibians generally there is complete crossing (decussation) of the optic nerves, right retina connecting to left tectum and vice versa. Since the mapping under consideration is from one surface to another, two axes are required to generate the ordered map which is obtained.

The technique used to establish this ordering does not resolve neighbourhoods less than a region containing about 1000 neurones on the tectum, due partly to the nature of the technique and partly to the fact that the map is not a neurone-to-neurone connection anyway: each retinal fibre branches extensively on the tectum and makes contact with many tectal neurones. However, it has been firmly established that the map is well ordered in two dimensions (actually in three, but only two will be considered here). Thus coordinate systems with conjugate 2-dimensional patterns of 'positional information' (Wolpert [7]) must exist over the retina and the tectum.

It has been assumed that the most plausible biochemical basis for an informational axis in an embryonic tissue is a monotonic gradient of some substance; but it is now evident that this is only one of a number of possible ways in which coordinate axes can be generated in embryos. The phase-shift model of Goodwin and Cohen [6], already described in earlier papers in this series (Cohen [1], Cooke and Goodwin [2]), provides an alternative to a substance gradient. This model has been applied in a rather general way to the retino-tectal projection in the original paper describing the theory. In the present study I will present a more detailed analysis geometrically speaking, while leaving out entirely any questions about biochemical details.

The starting point of the analysis is the normal map established by photic stimulation of the retina and neurophysiological determination of maximally sensitive response points in the tectum, shown in Figure 1. Here one sees the correspondence between regions in the left visual field and regions on the right optic tectum. Experiments which involved rotation of the embryonic eye

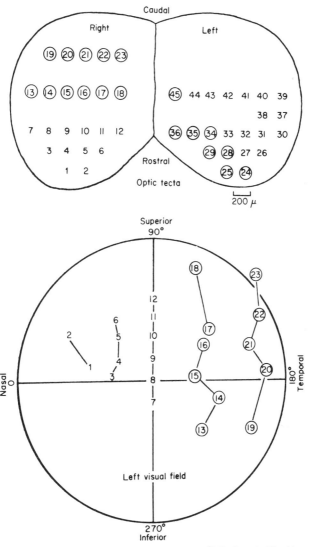

FIGURE 1. Normal map of the left visual field onto the right optic tectum in *Xenopus laevis* (from Gaze, Jacobson, and Szekely [4]).

through 90° and 180°, and construction of compound eyes by surgery and grafting, established the existence of two axes in the retina, each being determined at a characteristic time in embryonic development. Analogous experiments on the tectum are technically difficult, but the evidence points to a similar sequential determination of corresponding axes. Since these axes are roughly orthogonal, one running nasotemporally (i.e., head-to-tail) and the other dorso-ventrally in the retina, with correlative axes on the tectum (rostro-caudal and medio-lateral; i.e., head-to-tail and centre-to-side), it was natural to suggest that the pairs of gradients, whatever their nature, were also orthogonal, as depicted diagrammatically in Figure 2(a), (d). The maps for the two types of compound eye first studied, double nasal and double temporal (two half embryonic retinae joined with mirror symmetry along the vertical midline), showed some distortion of axes from an orthogonal relationship, but given the variation in the map from animal to animal, this did not seem to be sufficient to warrant any modification of the initial hypothesis.

Figure 3 shows the pattern obtained for the case of the double-nasal retina. The field is duplicated with mirror symmetry about the vertical midline, but the general characteristics of each map preserve the pattern of the normal projection (Figure 1). The gradient representation proposed for this compound eye is shown in Figure 2(b). Figure 2(c) gives the hypothetical gradient picture for the double-temporal retina.

It was not until double ventral eyes were studied that any serious discrepancy arose between observed and expected patterns. In this case the hypothesis of an orthogonal pair of gradients failed to generate the pattern obtained experimentally, and it became of interest to see if the patterns predicted by the phase-shift model would provide a better fit to observation. The

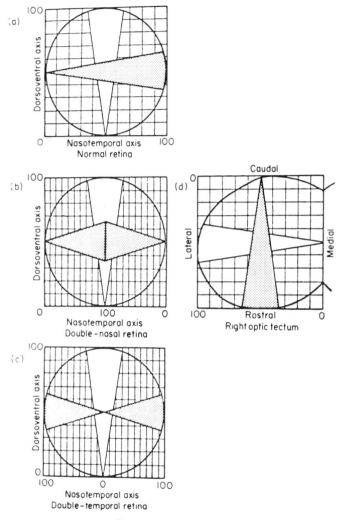

FIGURE 2. Proposed double gradient system in normal retina (a) and normal tectum (d) providing 2-dimensional grids of positional information over these surfaces. Co-ordinate systems proposed for compound eyes are as shown in (b) and (c) (after Gaze, Jacobson, and Szekely [4]).

determination was actually carried out 'blind': Dr. Gaze wisely withheld the unpublished experimental results from me until the map predicted by the phase-shift model had been drawn for the double ventral eye. The result was an unexpectedly good agreement between observation and prediction. This was particularly surprising in view of the number of degrees of freedom available in applying the model to any particular problem, and it suggested that there was some property of this model which is 'robust' in a general sense: it generates qualitatively correct contours, irrespective of details. This feature is a particularly simple one: the axes in the phase-shift model are not orthogonal because of an interaction of a particular kind between the axis-generating centres. This means of course that any model with the right kind of interaction would give the same result, so the successful prediction does no more than show that the phase-shift theory belongs naturally to the right class of model to provide an explanation for a particular phenomenon. A detailed description of the analysis will now be given.

Construction of the maps

In constructing a map over a two-dimensional surface, it is evident that the origins of the two coordinate axes must be located at points on the boundary of the surface if each point is to be labelled by a unique pair of coordinate values. Otherwise there will be a duplication of values or positional information, as occurs in compound eyes when the end of one axis lies at the centre of the surface (see Figure 2). Both the retina and the optic tectum are in fact curved surfaces, not plane as in Figure 2, and this curvature must be taken into account in constructing their projections onto a plane surface, as occurs when the actual mapping is performed experimentally. The retina is very nearly a section of a

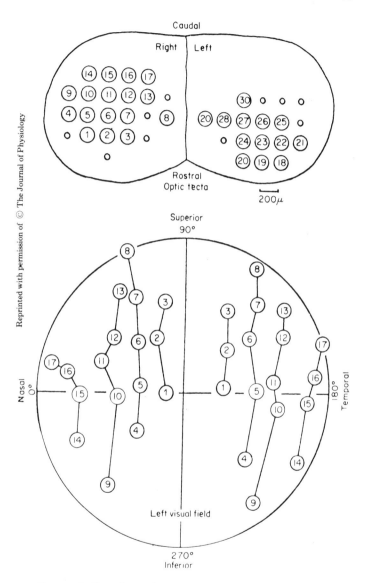

FIGURE 3. Experimental map of a left double-nasal retina onto the right optic tectum (after Gaze, Jacobson, and Szekely [4]).

sphere, but the tectum is more like a section of an ellipsoid. However, in the present analysis it has been assumed that both surfaces are sections of spheres. This introduces a distortion which would not be observable experimentally.

A comparison was made between the patterns of lines on the retina which would be expected to correspond to reference lines drawn on an orthogonal grid on the tectum (as in Figure 1) for two different models: (1) a double gradient model of the type assumed in Figure 2, each axis being determined by a linear gradient which varies independently of the other; and (2) the phase-shift model, which involves an interaction between the axes. This interaction arises because of the following considerations. It is assumed that one of the axes in the phase-shift model is generated from a pacemaker region of the tissue located at some point on the boundary. From this pacemaker region two activity waves, referred to as S and P_1, travel at different velocities, v and v_1 respectively, with $v > v_1$. Distance from the pacemaker, the origin of the axis, is measured by the time interval between the occurrence of the two events, as when one determines distance from a lightning flash by timing the interval between the arrival of the light and the sound waves. (This analogy was pointed out to me by Dr. D. Truman, to whom I am indebted for an illuminating metaphor.) Given constant propagation velocities over the tissue, a linear gradient of time or phase (since the waves recur periodically with a well-defined frequency) is produced.

The second axis is assumed to arise in consequence of a third propagating event, P_2, which originates in response to the S-event in a particular region of the tissue, again located at its boundary but in a region distinct from the pacemaker. This event propagates at velocity v_2, and distance along this axis is measured by the time interval (or phase angle) between S and P_2. Since these waves originate from different regions, the second coordinate axis will not be a straight line as in the double gradient model depicted in Figure 2, but will be curved, the curvature depending upon the relative velocities of v and v_2.

Since the observed axes in the retina are roughly naso-temporal (nose-to-tail) and dorso-ventral, it seemed most logical to locate the pacemaker and the P_2 propagation centres at the extremities of these axes. However, this still left eight possible allocations, choice among which could be dictated only by goodness of fit to the experimental material. This fit was established by eye, not by any quantitative measure, in view of the variability of the experimental results and the generally qualitative nature of the analysis. At this stage, only rather unexpected, counter intuitive results are significant, such as the double ventral mapping to be described below.

According to these criteria, the best choice was to locate the pacemaker at the temporal extremity of the retina and the P_2 propagation centre at the dorsal boundary. For matching axes in the tectum, the pacemaker is located at the rostral (most anterior) boundary and the P_2 propagation centre medially (in the midline). The nature of these events and the whole problem of how selective growth and recognition occurs in biochemical terms is left out of this analysis and will not be considered.

At this point it is necessary to add an embryological detail to the argument. When retinal fibres first make contact with the optic tectum, the embryo, now a tadpole, is about at stage 45 (a little more than four days old) and both retina and tectum are very much smaller than they are when the electrophysiological map is determined in the metamorphosed animal. However, the axes are laid down in these embryonic tissues and after determination remain specified throughout the life of the animal. In relation to the map this means that the growth process can be ignored. How this is possible in terms of actual biochemical and physiological processes is problematical, as discussed by Gaze [3], but this is a point that need not detain us in the present study.

Comparison of iso-informational contours according to the different models

Having selected the points of origin of the coordinate axes and idealized the surfaces to sections of spheres, the procedure for constructing the iso-informational contours (lines joining points with identical values along an axis) predicted by the two types of model under consideration for normal and compound eyes is straightforward. For the double gradient model, the contours are loci of points equidistant on the surface from the gradient origins. Since these origins are taken to be $\pi/2$ radians apart on the boundaries of the surfaces, the contours project to give essentially orthogonally intersecting curves. In constructing the map for compound eyes, the situation is slightly complicated by the problem of the size and the shape of the 'sources' for the substances producing the gradients. The experimental map for the case of a double nasal retina is shown in Figure 3. It is evident that medio-lateral lines on the tectum still map into dorso-ventral lines on the retina, as in the usual map (Figure 1). In order that these relationships be preserved in the model, it is necessary for the temporal source, which occupies a central position in the compound, double nasal retina (two half nasal retinae joined in mirror image fashion along the dorso-ventral midline) to be a line rather than a point, as

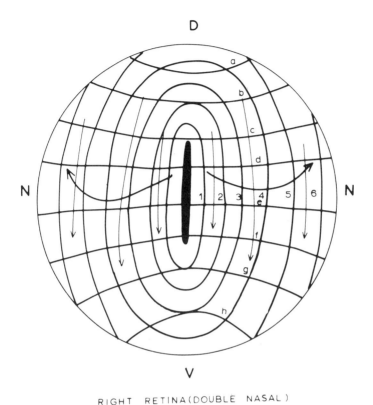

D

N   N

V

RIGHT RETINA(DOUBLE NASAL)

FIGURE 4. Hypothetical map of iso-informational contours on a double nasal right retina using the double gradient model and projecting the contours from a hemisphere onto a plane. One source of gradient substance is located dorsally, the other temporally, which becomes central in a double nasal retina.

shown in Figure 4. The contours labelled 1 to 6 are then elliptical rather than circular. This has the consequence that medio-lateral lines on the tectum map into the arrows running generally dorso-ventrally on the retina, three of which are shown on each half of the double nasal eye in Figure 4. The bold curved arrows in this figure, cutting across contours 1–5, correspond to the midline running rostro-caudally across the tectum, such as the line joining map positions 2, 6, 11, and 15 in Figure 3.

A point of possible confusion must now be clarified. It will be observed that all the experimental mapping results are presented in terms of points on an orthogonal grid on the tectum and corresponding points in the contralateral visual field (e.g., right tectum and left visual field in Figure 3). However, the models are all presented in terms of contours on the retina, as in Figures 2, 4, 5, 7–10. Because of the camera inversion effect of the lens of the eye, the relationship between visual field and retina is equivalent to a 180° rotation of either in relation to the other. Thus to convert the positions of visual field points shown in Figure 3 into points on the left retina (projected onto a plane), one simply rotates the figure through 180° and relabels the reference positions so that inferior field becomes dorsal retina, superior field becomes ventral retina, temporal becomes nasal and nasal, temporal. The theoretical maps were all constructed for the right retina and the left tectum, because this was the experimental map first requested for the test of the model in the case of the double ventral eye.

Since the right-left inversion has no effect on the double nasal eye, in view of its symmetry about the dorso-ventral axis, one can simply invert Figure 3 in order to make the comparison with Figure 4. Unfortunately the figures are not detachable.

Although the correspondence between lines on the model and those observed is not by any means highly accurate, it is perfectly satisfactory within the degree of reproducibility attainable in the experiments.

At this point let us compare this correspondence with that anticipated on the basis of the phase-shift

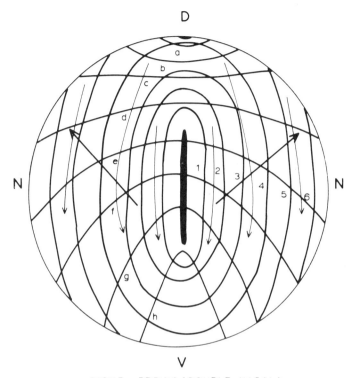

FIGURE 5. Hypothetical map of iso-informational contours on a double nasal right retina using the phase-shift model, assuming a temporal pacemaker and a dorsal P_2-propagation centre.

model. The phase contours are as shown in Figure 5. The three lines running dorso-ventrally in each half retina correspond to the light lines running latero-medially across the tectum, as shown in Figure 6, while the heavy arrow running diagonally across the half retinae correspond to the rostro-caudal arrow on the tectum. The construction procedure used to establish these contours is as follows. Taking both retina and

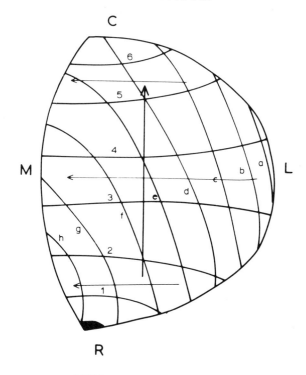

LEFT TECTUM

FIGURE 6. Phase contours on the left tectum according to the phase-shift model, assuming a rostral pacemaker and a lateral P_2 propagation centre. Reference arrows are drawn orthogonally on this surface.

tectum to be sections of spheres, and assigning the pacemaker and P_2-propagation centres to appropriate points on the boundaries of these sections, contours of constant phase angle $\phi_{P_1 S}$ are constructed as lines at constant distance on the spherical surface from the pacemaker centre. Figure 7 shows the procedure for the right retina, where the point P is the pacemaker centre located at the temporal extremity of the retina. The point Q is the dorsal extremity, the P_2 propagation

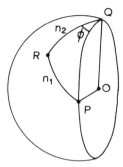

FIGURE 7. Construction for determining phase contours
on the basis of the phase-shift model. Details as in text.

centre, while O would correspond to the centre of the
lens. The retina is thus taken to be a hemisphere, the
angle QOP being a right angle. R is a point on the surface.

To construct the contours of constant phase
ϕ_{P_1S} one joins points R for which the solid angle n_1 is
constant. This gives the curves marked 1–6 in Figure 8
where the hemisphere is projected onto a plane. The
corresponding contours on the tectum are labelled
1–6 in Figure 6. These curves are identical with those
obtained on the assumption that P is a source of
chemical substance, the contours then joining points
of equal concentration.

The other contours in the phase-shift model are
obtained by using the following pair of equations defining the relationship between n_1 and n_2, the solid
angle distances of R from points P and Q respectively:

(1) $$2n_2 - n_1 = \text{const},$$

(2) $$\cos n_2 = \cos \phi \cos(\pi/2 - n_1).$$

Equation (1) arises from the assumption that the S
event propagates at twice the velocity of the P_2 event,
and equation (2) from the constraint that the points R

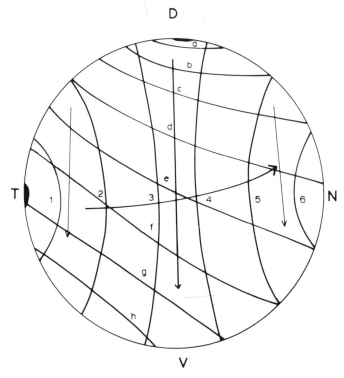

RIGHT RETINA (NORMAL)

FIGURE 8. Phase contours on the right retina.

lie on a spherical surface, with P and Q $\pi/2$ radians apart
on this surface. The angle ϕ is as shown in Figure 7.
Choosing some value for the constant in equation (1),
one can then find the values for one of the variables,
say n_1, by solving for n_2 from (1), substituting in (2),
and allowing ϕ to take values between 0 and π. The
corresponding values for n_2 can then be determined.
Each value assigned to the constant in (1) defines a
different contour of constant phase difference, ϕ_{P_2S}.
These contours are labelled a to h in Figure 8 for the
retina, in Figure 6 for the tectum, and in Figure 5

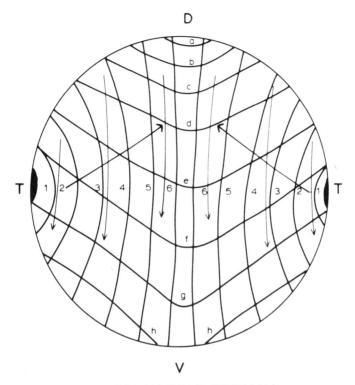

RIGHT   RETINA (DOUBLE   TEMPORAL)

FIGURE 9. Phase contours on a double temporal right retina.

for the double nasal retina. Comparing Figure 5 with Figure 4 one sees that the effect of the interaction in the phase-shift model is to cause a progressive skewing of the contours a to h as one moves ventrally so that symmetry of intersection angles about the horizontal midline is not preserved. This skewing is also evident in Figures 6 and 8. The difference in the maps derived from the two models for the double nasal retina is simply a difference of curvature in the arrows corresponding to the rostro-caudal axis in the tectum (heavy black arrows, Figures 4 and 5). The experimental

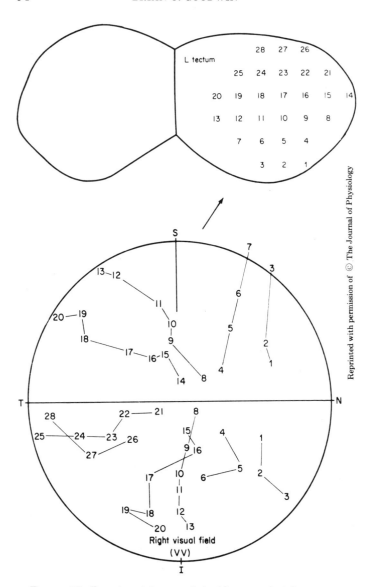

FIGURE 10. Experimental map of double ventral right eye onto the left optic tectum (after Gaze, Keating, and Straznicky [5]).

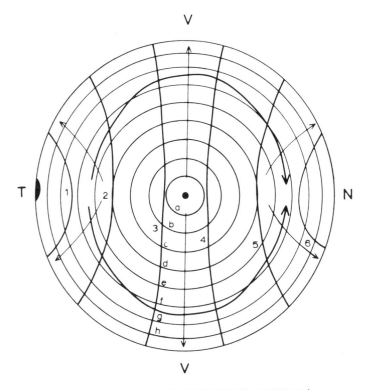

RIGHT   RETINA(DOUBLE   VENTRAL)

FIGURE 11. Contours on a double ventral right retina according to a double gradient model.

map, Figure 3, gives a picture intermediate between these.

Turning next to the double temporal retina, the picture given by the phase-shift model is shown in Figure 9. As with the double nasal retina, the two models do not give significantly different results, and so neither the alternative model nor the experimental map is shown.

The experimental map for the test case, the double ventral retina, is reproduced in Figure 10, which shows

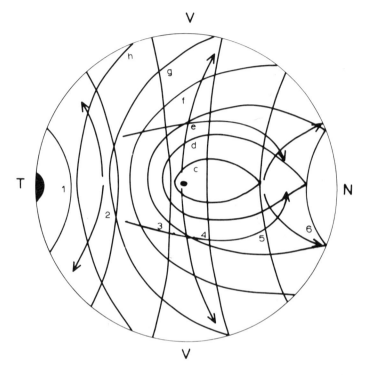

RIGHT RETINA(DOUBLE VENTRAL)

FIGURE 12. Phase contours on a double ventral right retina according to the phase-shift model.

the map of the right visual field. The major feature of this map is the absence of symmetry about the vertical midline: the most nasal lines are nearly vertical, becoming progressively more horizontal as one moves in a temporal direction. This type of asymmetry cannot be generated by a double-gradient model in which there is no interaction between the sources. The predicted map for the double ventral eye on the basis of this kind of model is shown in Figure 11. The central source is represented as a point, but changing its shape would not alter the symmetry properties of the map. Lines corresponding to the usual orthogonal reference lines on the tectum are as shown, and do not give any satisfactory degree of correspondence with the experimental picture.

On the other hand, the phase-shift model gives a map which is in reasonably good agreement with the observations, as shown in Figure 12. Lines running latero-medially across the tectum map into lines which run more or less dorso-ventrally across the retina in its temporal region (corresponding to nasal visual field), these lines becoming more and more curved towards the horizontal as one moves nasally across the retina (temporally across the visual field). The other axis also fits the experimental picture, whereas the corresponding axis for the double gradient model does not.

Conclusions

The analysis presented above, being qualitative, cannot be more than suggestive of features which must be incorporated into any model which sets out to explain some of the observations on the retino-tectal projection in amphibia. The experimental material in this field is exceedingly elegant and has a degree of resolution greater than any other area of field analysis in embryology, exploiting as it does neurophysiological techniques. However, it still falls short of the accuracy that would be required to choose with confidence between alternative models of the type considered here. Thus the analysis given in this paper must be treated as no more than an attempt to develop a theoretical picture providing a set of analytical procedures which can be used in conjunction with experimental techniques whose sophistication and resolving power are increasing all the time. Much more direct experimental tests are necessary for making choices between the phase-shift and the source-diffusion models of embryonic field formation. However, it is evident that experimental embryology has now advanced to the point where theoretical analyses can be of some use in clarifying unexpected observations in certain areas, and can also be of some value in designing further experiments.

Acknowledgement. I would like to acknowledge the generosity of the Science Research Council in providing assistance for research in Theoretical Biology.

REFERENCES

1. M. H. Cohen, *Models of clocks and maps in developing organisms,* Lectures on Mathematics in the Life Sciences, vol. 3, Amer. Math. Soc., Providence, R. I., 1971.

2. J. Cooke and B. C. Goodwin, *Periodic wave propagation and pattern formation: Application to problems in development,* Lectures on Mathematics in the Life Sciences, vol. 3, Amer. Math. Soc., Providence, R. I., 1971.

3. R. M. Gaze, *The formation of nerve connections,* Academic Press, New York and London, 1970.

4. R. M. Gaze, M. Jacobson and G. Szekely, J. Physiol. **165** (1963), 484.

5. R. M. Gaze, M. J. Keating and A. Straznicky, J. Physiol. **214** (1971), 37.

6. B. C. Goodwin and M. H. Cohen, J. Theoret. Biol. **25** (1969), 49.

7. L. Wolpert, J. Theoret. Biol. **25** (1969), 1.

SOME STATISTICAL ASPECTS OF THE THEORY
OF INTERACTING SPECIES

By

ELLIOTT W. MONTROLL

University of Rochester

Introduction

Recently N. S. Goel, S. C. Maitra, and the author [1] made a detailed analysis of the nonlinear Lotka-Volterra model of interaction between species; the same model which has been discussed in the previous lecture by Professor Kerner [2]. This paper is an elaboration of a number of ideas in [1]. The most novel part of the work to be reported is the development of a nonlinear model similar to that of Lotka and Volterra, but whose associated sets of coupled differential equations can be solved exactly by employing what the author has called the cheap trick [3] of making a transformation in the dependent variables. This transformation reduces the equations to linear ones which can be solved by standard methods. Most of the concepts discussed by Kerner in connection with the Lotka-Volterra model; the existence of a constant of the motion [4], the possibility of constructing a canonical ensemble [2], etc., will also be apparent in the new model.

1. Models of growth of a population to saturation

It was observed by Malthus [5] many years ago that the population of Europe seemed to be doubling at regular time intervals, a characteristic of an exponential increase. The rate of population increase in such circumstances is proportional to the population itself

$$(1.1) \qquad dN/dt = kN.$$

The population at a time t, $N(t)$ in a given region is then

$$(1.2) \qquad N(t) = N(0) \exp kt,$$

By the 1830's it was noted by several people that the

strength of the exponentiation process seemed to be declining. Quetelet suggested that saturation might even be expected and his student Verhulst [6] produced a correction factor to the Malthus equation (1.1) which would imply saturation

$$(1.3) \qquad dN/dt = kN[1 - (N/\theta)]$$

where θ is the saturation population. Of course, as $\theta \to \infty$ this reduces to (1.1).

The solution of (1.3) is called the logistic curve and has the form

$$(1.4) \quad N(t) = \theta N(0)/\{N(0) + [\theta - N(0)]\exp(-kt)\}.$$

Census data of a number of countries can be fitted quite well by this equation. The U. S. curve [3] whose three parameters $N(0)$, k, and θ were chosen to fit census data exactly at 1840, 1900, and 1960 is in Figure 1. The points on the curve represent the census since 1800.

During the early 1920's, Pearl and Reed [7] analyzed the population growth of many European countries by employing the form

$$(1.5) \qquad N(t) = C_0 + C_1/[1 + C_2 e^{-\alpha t}].$$

The extra parameter C_0 was introduced so that two growth regimes could be characterized by one equation. The European population from about 1200 to 1700 fluctuated somewhat around a mean value that did not rise significantly until the exponentiation started in the early 18th century. The constant C_0 represents that mean value so that if $t = 0$ corresponds to 1800, C_0 is the value of $N(t)$ at $t = -\infty$. Equation (1.5) gave a remarkably accurate description of the population data of many countries. Two striking exceptions [8] were Ireland (with population drop from 8 million in 1840 to 4 million by 1940 following the potato famine of 1845 and a massive emigration and drop in birth rate

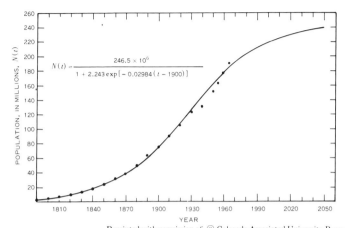

$$N(t) = \frac{246.5 \times 10^6}{1 + 2.243 \exp[-0.02984(t - 1900)]}$$

Reprinted with permission of © Colorado Associated University Press

FIGURE 1. Population of U. S. logistic curve fitted so that observed points at 1840, 1900 and 1960 are exact. Points represent census data.

in the period 1850 − 1940) and The Netherlands (whose size was changed by reclaiming land from the sea). Most of the countries which retained their identity with little change after World War II followed population trends predicted by Pearl [7]. His 1960 predictions (in millions) made in 1924 are compared with 1960 census data as given below:

	Pearl	Census 1960
Sweden	6.9	7.5
Belgium	11.0	9.1
Denmark	4.4	4.6
England and Wales (1961)	52.1	46.1
France	41.4	46.5
Norway	3.2	3.6
Scotland	6.2	5.2

Pearl's 1924 prediction for the sum of these populations 125.2×10^6 exceeds the census report figure of 122.6×10^6 by only 2.6×10^6.

The census deviations from the logistic curve as seen in Figure 1 for the depression period of $1930-1940$ and for the post-World War II period require a separate discussion which will be the subject of the next section.

One of the features of the logistic law which is very special and which might cause one to question the law for fitting purposes is the fact that it has a certain symmetry about the inflection point so that the deviation of the logistic from saturation at points above the inflection point has the same form as the growth above zero at points below the inflection point. A family of empirical curves which saturates either more slowly or more rapidly than the initial form of the rise can be derived from the following generalization of the Verhulst equation (1.3):

(1.6) $$dN/dt = kN[1 - (N/\theta)^{\alpha}]/\alpha,$$

which is the same as (1.3) when $\alpha = 1$. The special case $\alpha = 0$,

(1.7) $$dN/dt = -kN\log(N/\theta),$$

is known in the population literature as the Gompertz equation [9]. It was first proposed (with $k < 0$) in connection with mortality analysis of elderly people, an application which employs a different interpretation of N. dN/dt represents the rate at which persons who were born in a given year die as a function of time. The use of (1.7) for the analysis of population curves is discussed in reference [10]. It never became popular because, in the precomputer period, its solution, as will be given below, was more difficult to fit by least squares to census data than was that of the Verhulst equation.

The solution of (1.6) is obtained by making the transformation

(1.8a) $$v = \log(N/\theta).$$

Then (1.6) is equivalent to

(1.8b) $$d \log(1 - e^{-\alpha v}) = -d(kt)$$

or

(1.8c) $$\left\{ \frac{1 - \exp[-\alpha v(t)]}{1 - \exp[-\alpha v(0)]} \right\} = \exp(-kt).$$

When $\alpha = 0$ this becomes $[v(t)/v(0)] = \exp(-kt)$, or

(1.9) $$N(t) = \theta \{ N(0)/\theta \}^{\exp(-kt)}.$$

For other values of α a more useful form of the solution of (1.6) is

(1.10) $$N(t) = \theta \left\{ \frac{[N(0)]^\alpha}{[N(0)]^\alpha (1 - e^{-kt}) + \theta^\alpha e^{-kt}} \right\}^{1/2}$$

which reduces to (1.4) when $\alpha = 1$.

2. Population variation under random influences

We have already alluded to the deviations of the U. S. population curve from the logistic in Figure 1. In various experiments on animal populations it is noted that as the population in a limited volume with limited resources approaches saturation, some random oscillations about the saturation level are observed. The sheep population in Tasmania, as exhibited in Figure 2, shows such a variation also. These fluctuations reflect changes in the birth and death rates. The number of sheep grown depends on the economy, as does the number of people born. The U. S. birth and death rates since 1900 are plotted in Figure 3. Note the drop in birth rate during the depression and the rise in the optimistic post-war boom of the 1950's. Epidemics, especially those in The Black Plague, contributed significantly to the population fluctuations of the Middle Ages.

Some influences on the birth rate might be changes in the economy which stimulate or oppose interest in marrying early and having children, changes in average size of dwellings, changes in public attitudes (for example, in one year it is stated to be essential that high IQ people should have more children, in another year the zero population growth movement is popular; during one year birth control pills are reputed to be safe, the next year they are dangerous, etc.), variation in government tax regulation and social security policies, etc. If the U. S. birth rate curve continues to drop as it has for the past ten years, the death and birth rate levels will be about the same in the middle 1980's, a state which already exists in some countries in the Soviet block.

In view of these fluctuating population driving forces, it would seem that a more appropriate model for population variation would be one characterized by the equation

$$(2.1) \qquad dN/dt = kNG(N/\theta) + NF(t)$$

where $F(t)$ is considered to be a random function of time which reflects changes in the economy and attitudes as described above. Since it is very hard to predict changes in attitude in detail, one of the few mathematical approaches left open to us is to incorporate them all in the random function $F(t)$. The form of $G(x)$ which was discussed in the last section was

$$(2.2) \qquad G(x) = (1 - x^\alpha)/\alpha,$$

$\alpha = 0$ being the Gompertz case and $\alpha = 1$ the Verhulst case.

Several postulates will be made about the statistical properties of $F(t)$ and their consequences will be examined. We assume first that $F(t)$ is generated by a Gaussian random process. Then we assume that the average value of $F(t)$, $\langle F(t) \rangle$, vanishes,

$$(2.3) \qquad \langle F(t) \rangle = 0,$$

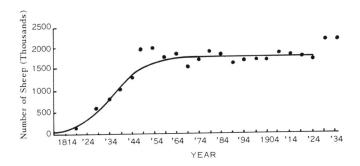

FIGURE 2. Population growth of sheep introduced into Tasmania. The dots represent average numbers over five-year periods [from J. Davidson, in Transactions of the Royal Society of South Australia, 62:342-346, 1938].

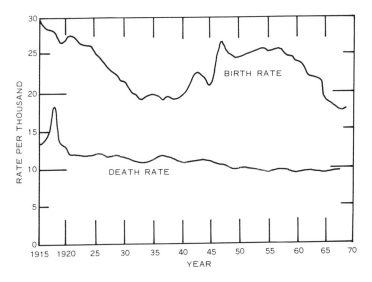

FIGURE 3. Variation of birth and death rates per thousand per year in the period 1915-1970 (data from [10]).

while the average value of the correlation of $F(t_1)$ and $F(t_2)$ is proportional to a delta function as usually postulated in the theory of Brownian Motion

$$(2.4) \qquad \langle F(t_1) F(t_2) \rangle = \sigma^2 \delta(t_1 - t_2).$$

The random variable $F(t)$ is then characterized by a single parameter σ.

If $\langle F(t) \rangle \neq 0$, then in most interesting cases, either the saturation level or the rate constant can be changed so that the average value of the random component vanishes. Let us suppose that

$$(2.5) \qquad F(t) = a + \delta F \quad \text{and} \quad \langle \delta F \rangle = 0.$$

Then for the form (2.2) of $G(x)$,

$$(2.6) \qquad \begin{aligned} dN/dt &= kN[1 - (N/\theta)^\alpha]/\alpha + aN + N\delta F \\ &= k'N[1 - (N/\theta')^\alpha]/\alpha + N\delta F \end{aligned}$$

where

$$(2.7a) \qquad k' = (k + \alpha a),$$

$$(2.7b) \qquad \theta' = \theta\{1 + (\alpha a/k)\}^{1/\alpha};$$

as $\alpha \to 0$, $\theta' = \theta \exp(a/k)$.

The postulate (2.4) is appropriate when the random forcing function $F(t)$ has a memory which is very short compared with the life expectancy of an individual or with the period between generations. This is generally the case when economic and popular responses of the type mentioned earlier in this section are the basis of the driving force. W. L. Thorp and W. C. Mitchell [11] have analyzed the economic conditions of fifteen countries during the period $1790 - 1925$. The duration of a business cycle was defined as the time interval between one peak in the prosperity of a country and the next one, a relative recession occurring between two successive

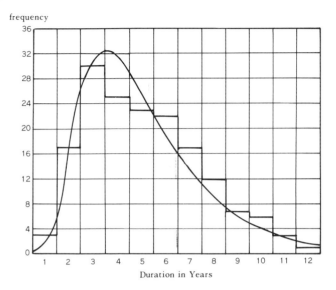

FIGURE 4. Logarithmic normal fit to the frequency distribution of the observed duration of business cycles (see [11]).

peaks. Thorp and Mitchell then found the frequency distribution of duration of business cycles to be a log normal distribution which peaked for cycles between 3 and $3\frac{1}{2}$ years (see Figure 4), a time short compared with the life expectancy of an individual. Various fads and popularist movements generally seem to last a similar length of time.

If we again let

$$(2.8) \qquad v = \log N/\theta,$$

equation (2.1) becomes

$$(2.9) \qquad dv/dt = G(e^v) + F(t).$$

This equation and the various postulates made above are just those required to derive a Fokker-Planck equation [12] for the probability that the $\log N/\theta$ has a value v at time t, $P(v, t)$. It has been shown in [1] that the Fokker-Planck equation has the form

$$(2.10) \qquad \frac{\partial P}{\partial t} = - k \frac{\partial}{\partial v} \{ PG(e^v) \} + \tfrac{1}{2}\sigma^2 \frac{\partial^2 P}{\partial v^2}.$$

The equilibrium distribution function can be obtained by setting $\partial P/\partial t = 0$. Then it must satisfy

$$- k \frac{\partial}{\partial v} \{ PG(e^v) - (\sigma^2/2k) \partial P/\partial v \} = 0.$$

That is, with the normalization constant P_0,

$$(2.11) \qquad P(v, \infty) = P_0 \left\{ \exp(2k/\sigma^2) \int_0^v G(e^v) \, dv \right\}.$$

In particular, when

$$(2.12a) \qquad\qquad G(e^v) = (1 - e^{\alpha v})/\alpha,$$

$$(2.12b) \ P(v, \infty) = P_0 \exp\{ (2k/\alpha^2\sigma^2)(1 + \alpha v - e^{\alpha v}) \}.$$

The easiest case to deal with is the Gompertz form of $G(e^v)$, that with $\alpha = 0$, for then $P(v, \infty)$ becomes Gaussian. The normalized equilibrium population distribution in terms of N is then

(Gompertz case)

$$(2.13) \ P(N, \infty) = (k/2\pi\sigma^2)^{1/2} N^{-1} \exp\{ - k[\log(N/\theta)]^2/2\sigma^2 \}.$$

The general normalized distribution function in terms of N is.

$$P(N, \infty) = \frac{\alpha}{\theta} \left\{ \frac{2k}{\alpha^2\sigma^2} \left(\frac{N}{\theta} \right)^\alpha \right\}^{2k/\alpha^2\sigma^2} \left\{ \frac{(2k/\alpha^2\sigma^2)^{1/\alpha}}{\Gamma(\alpha^{-1} + [2k/\alpha^2\sigma^2])} \right\}$$

$$(2.14) \qquad\qquad\qquad\qquad \cdot \exp \left\{ - \frac{2k}{\alpha^2\sigma^2} \left(\frac{N}{\theta} \right)^\alpha \right\}.$$

The equilibrium distribution function (2.12b) has the interesting property that if v is small (i.e., if deviations from the population saturation level are small), then $P(v, \infty)$ is independent of α and has the Gaussian form

(2.15) $P(v, \infty) = P_0 \exp\{ - kv^2/\sigma^2 \}.$

An alternative form of the Fokker-Planck equation of our process is obtained by letting [1]

(2.16) $P(v,t) = \Psi(v,t) \exp \left\{ k\sigma^{-2} \int_0^v G(e^v) \, dv \right\}$

then

(2.17) $(2/k)\Psi_t = \sigma^2 k^{-1}\Psi_{vv} - \left\{ \dfrac{\partial G(e^v)}{\partial v} + k\sigma^{-2}[G(e^v)]^2 \right\} \Psi.$

This is to be compared with the Schrödinger equation

(2.18) $\hbar i \Psi = (\hbar^2/2m)\Psi_{xx} - U(x)\Psi$

and the Bloch equation in which $- it/\hbar$ is replaced by $\beta = 1/kT$ (which is used in statistical mechanics),

(2.19) $\Psi_\beta = (\hbar^2/2m)\Psi_{xx} - U(x)\Psi.$

Notice that if, in the Bloch equation, we choose the mass to be $\frac{1}{2}$ and then identify β with $\frac{1}{2}kt$, $\hbar^2$ with σ^2/k, and $U(x)$ with

(2.20) $W(v) \equiv k\sigma^{-2}[G(e^v)]^2 + \partial G(e^v)/\partial v,$

it has the same form as our basic equation (2.17). Of course, there is no connection between the physical significance of the two equations; however, there is a mathematical convenience in their similarity because the literature on the Schrödinger and Bloch equations is immediately available to us.

Let us again choose the special form of $G(e^v)$ (2.12a). Then

(2.21) $W_\alpha v \equiv k(\alpha\sigma)^{-2}(1 - e^{\alpha v})^2 - e^{\alpha v}.$

In the Gompertz $\alpha = 0$ case,

(2.22) $W_0(v) = (k/\sigma^2)v^2 - 1,$

which is the harmonic oscillator potential in quantum theory. Generally

(2.23a) $W_\alpha(v) = A(e^{2x\alpha} - 2e^{x\alpha}) + k(\alpha\sigma)^{-2}$

where

(2.23b) $x \equiv (v - v^*)$, $\exp(\alpha v^*) = 1 + [(\alpha\sigma)^2/2k]$

(2.23c) $A \equiv [k/(\alpha\sigma)^2]\{1 + [(\alpha\sigma)^2/2k]\}^2$.

If we introduce a new function Φ by

(2.24) $\Psi = \Phi \exp\{-\tfrac{1}{2}k[E + k(\alpha\sigma)^{-1}]t\}$

then Φ satisfies the differential equation

(2.25) $(\sigma^2/k)\Phi_{xx} + \{E - A(e^{2\alpha x} - 2e^{\alpha x})\}\Phi = 0$

which is just the Schrödinger equation for a diatomic molecule with a Morse potential [13] when the reduced mass is taken to be $\tfrac{1}{2}$ and $\hbar^2$ is identified with σ^2/k. Generally, our x is replaced by $-x$ in studying diatomic molecules. Mathematically this difference is of no importance. We seek solutions of (2.25) which vanish at $x = \pm \infty$.

Discussions of the Schrödinger equation for the harmonic oscillator and the Morse potential are available in the quantum mechanics literature. The relevant results were translated into the language required for our equations in [1]. We merely state some of the results here. The Gompertz $\alpha = 0$ case which corresponds to a harmonic oscillator is easiest to analyze. One finds that if at time $t = 0$, the population is $n(0) = n_0$ so that

(2.26) $v_0 = \log(N_0/\theta)$;

then the probability of a transition from v_0 to v in time t (i.e., of a population transition from $N(0) = \theta \exp v_0$ to $N(t) = \theta \exp v$) is

$$P(v, v_0; t) = \left\{\frac{k}{\pi\sigma^2(1 - e^{-2kt})}\right\}^{1/2}$$

(2.27) $\cdot \exp\{-(v - v_0 e^{-kt})^2 k/\sigma^2(1 - e^{-2kt})\}$.

The variation of various moments of the population can be determined as a function of time from the above equation [1]:

(2.28a) $\langle N/\theta \rangle = (N_0/\theta)^{\exp(-kt)} \exp[(\sigma^2/4k)(1 - e^{-2kt})]$

while

(2.28b)$\langle (N - \overline{N})^2 \rangle / (\overline{N})^2 = -1 + \exp\{(\sigma^2/2k)(1 - e^{-2kt})\}.$

Generally

(2.28c) $\langle (N/\theta)^{2\lambda} \rangle = (N_0/\theta)^{2\lambda \exp(-kt)} \exp\{(\lambda \sigma^2/k)(1 - e^{-2kt})\}.$

An arbitrary initial distribution develops according to

(2.29) $P(v, t) = \int_{-\infty}^{\infty} P(v, v_0; t) P(v_0, 0) \, dv_0.$

We will not discuss the general saturation inducing function $G(e^v)$ in detail, but we will note its consequences in two important regimes. The first is the regime far from saturation. Let us assume that as $\theta \to \infty$,

(2.30) $G(N/\theta) \to 1/\alpha < \infty$ and $k' = k/\alpha.$

Then the Fokker-Planck equation (2.10) becomes

(2.31) $\dfrac{\partial P}{\partial t} = -k' \dfrac{\partial P}{\partial v} + \sigma^2 \dfrac{\partial^2 P}{\partial v^2}$

whose solution is [1]

$P(v, t) = (2t\pi\sigma^2)^{-1/2} \int_{-\infty}^{\infty} P(v', 0)$

(2.32) $\cdot \exp\{-(v - v' - k't)^2/2t\sigma^2\} dv'.$

In the case that the population is precisely N_0 at time $t = 0$,

(2.33) $P(v', 0) = \delta(v' - v_0)$

where $v' - v_0 = \log N'/N_0$ or $N'/N_0 = \exp(v' - v_0)$. Then the probability that v lies between v and $v + dv$ is

$$P(v,t)\,dv = (2t\pi\sigma^2)^{-1/2}\exp\{-(v - v_0 - k't)^2/2t\sigma^2\}\,dv,$$

(2.34) $-\infty < v < \infty,$

so that, as $\sigma \to 0$, v follows the Malthusian exponential trajectory

(2.35) $v - v_0 = tk'$ or $N/N_0 = \exp tk'.$

The probability that N lies between N and $N + dN$ at time t is

$$P(N,t)\,dN = \frac{dN\exp\{-(\log[(N/N_0)e^{-tk'}])^2/2t\sigma^2\}}{N(2t\pi\sigma^2)^{1/2}},$$

(2.36) $0 < N < \infty.$

The first two moments of this distribution are

(2.37a) $\overline{N} = N_0\exp(k' + \tfrac{1}{2}\sigma^2)t,$

(2.37b) $\langle(N - \overline{N})^2\rangle/N^2 = -1 + \exp t\sigma^2.$

The next important regime is that in which v is small. This means the system is making small fluctuations about saturation. That is exactly the equilibrium distribution (2.11). When v is small, (since $G(1) = 0$),

(2.38a) $G(e^v) = vG'(1) + \cdots$

so that

(2.38b) $P(v, \infty) \cong P_0\{\exp[(ev^2/\sigma^2)G'(1) + O(v^3)]\}.$

Generally $G'(1) < 0$. For example, when (2.12a) is chosen for $G(e^v)$,

(2.38c) $G'(1) = -1.$

The intermediate regime in which the finiteness of θ becomes apparent, but before saturation occurs, depends on the detailed form of $G(x)$. Some of these details are given in [1] for the form (2.12a).

In conclusion we summarize our findings for the form (2.12a) as follows: In the first (the $\theta \to \infty$) regime the population grows freely with no interference. This is analogous to a free particle which accelerates in a field. In the second regime, the population has grown to the point that it is affected by other influences such as other species (and, in the case of human population growth, by fluctuations in the economy, by changes in personal attitudes, by agricultural successes and failures, etc.). In our Morse-type equation, this is analogous to the system falling into the highest energy bound state of the Morse potential, then dropping into lower energy states until it reaches the ground state. In the ground state the population fluctuates around its average value with statistics characterized by the equilibrium distribution (2.38b). These fluctuations are the analogues of the zero point fluctuations of a Morse oscillator.

3. Models of interacting species

If we interpret the population $N(t)$ discussed in the previous section as the population of some animal species living in the forest or in the ocean, the random function $F(t)$ of equation (2.1) might be considered to be the influence of other species which survive by eating the one of interest (a negative contribution) as well as the variation of food supply for it. Some populations of wild species seem to have a periodicity associated with them; there are even pairs of species whose populations seem to vary with the same frequency. A classical example is the periodic variation of populations of the lynx and the snowshoe hare [14] as indicated in Figure 5. Apparently the lynx prey on the hare so that as the population of the lynx increases, that of the hare decreases, until there are not enough hares to support the elevated population of lynx thus causing some of the lynx to starve. With less lynx, the hare population again increases, etc.

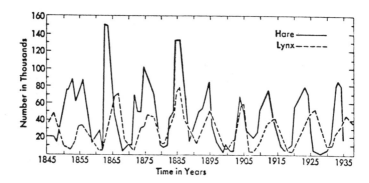

FIGURE 5. Changes in the abundance of lynx and snowshoe hare [from D. A. MacLulich, Univ. of Toronto Studies, Bio. Series 43, 1937].

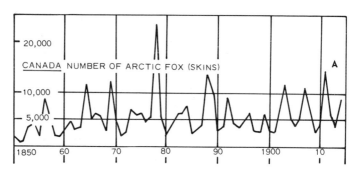

FIGURE 6. Number of Arctic fox skins taken annually by the Hudson Bay Company [from C. S. Elton, J. Exp. Biol. 2 (1924), 119–163.].

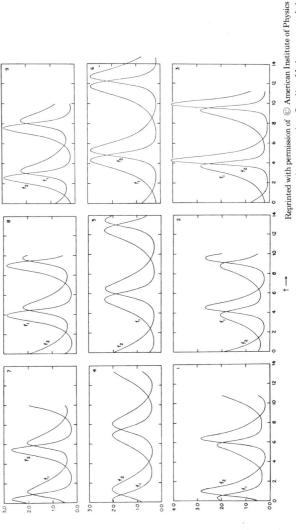

FIGURE 7. Time variation of population of two species interacting according to the Lodka-Volterra model. The values of k_1 and k_2 are for the indicated figures (1), (2), (3) $k_1 = 1$, $k_2 = 2$; (4), (5), (6) $k_1 = k_2 = 1$ and (7), (8), (9) $k_1 = 2$, $k_2 = 1$. The initial values of t_1 and t_2 are those given for $t = 0$ on the graphs (from [1]).

Reprinted with permission of © American Institute of Physics

Other examples have also been observed. Figure 6 is a record of the number of arctic fox skins taken by the Hudson Bay Company. D'Ancona [15] made a statistical analysis of Adriatic fish catches in the Adriatic during the period 1905 − 1923 and found similar variations in the population of certain species of fish.

The classical theory of two interacting species is that due to Lotka [16] and Volterra [4]. It is assumed that one of the species (which we call species 1) would grow exponentially in the absence of the second species (species 2) if it were not preyed upon by it. It is also postulated that species 2 would die out with a rate constant k_2 in the absence of species 1. The Lotka-Volterra equations are then

$$(3.1a) \qquad dN_1/dt = k_1 N_1 - \lambda_1 N_1 N_2,$$

$$(3.1b) \qquad dN_2/dt = - k_2 N_2 + \lambda_2 N_1 N_2.$$

The constant λ_1 tells how rapidly species 1 would die out through encounters with species 2, and λ_2 is the rate constant for the increase in species 2 due to encounters with 1.

It is convenient to define

$$(3.2) \qquad f_1(t) \equiv \lambda_2 N_1(t)/k_2 \quad \text{and} \quad f_2(t) \equiv \lambda_1 N_2(t)/k_1$$

which satisfy the pair of equations

$$(3.3a) \qquad df_1/dt = k_1 f_1(1 - f_2),$$

$$(3.3b) \qquad df_2/dt = - k_2 f_2(1 - f_1).$$

By eliminating the time we find the constant of the motion

$$(3.4) \qquad (f_1 e^{-f_1})^{1/k_1} (f_2 e^{-f_2})^{1/k_2} = \text{constant}.$$

From this relation it can be deduced that f_1 and f_2 are periodic functions of the time [4], [3]. We have plotted the variation of f_1 and f_2 in Figure 7, starting with various initial values of f_1 and f_2. If these are close to the equilibrium values $f_1 = f_2 = 1$, the time dependence of both f's is sinusoidal. When the initial points are far from equilibrium, the periodic curves tend to be more spiked and resemble observed population variations more closely. Volterra [4], [15] generalized the equations (3.1) to a set which describe the interaction of n species

$$(3.5) \qquad \frac{dN_i}{dt} = k_i N_i + \beta_i^{-1} \sum_{j=1}^{n} a_{ij} N_i N_j.$$

The first term describes the behavior of ith species in the absence of others; when $k_i > 0$, the ith species is postulated to grow in an exponential Malthusian manner, with k_i as the "rate constant." When $k_i < 0$ and all other $N_j = 0$, the population of the ith species would die out exponentially. The quadratic terms in equation (3.5) describe the interaction of the ith species with all the other species. The ith term in the quadratic sum is proportional to the number of possible binary encounters $N_i N_j$ between members of the ith species and members of the jth species. The constants a_{ij} might be either positive, negative, or vanish. A positive a_{ij} tells us how rapidly encounters between ith and jth species will lead to an increase in N_i; a negative a_{ij} tells how rapidly these encounters will lead to a decrease in N_i, and a zero a_{ij} simply denotes the fact that ith and jth species do not interact. If during a collision between ith and jth species, jth species are gained, then ith species are lost. Hence a_{ij} and a_{ji} have opposite signs. The positive quantities β_j^{-1} have been named "equivalence" numbers by Volterra. During binary collisions of species i and j, the ratio of i's lost (or gained) per unit time to j's gained (or lost) is $\beta_i^{-1}/\beta_j^{-1}$. With this definition,

$$(3.6) \qquad a_{ij} = -a_{ji}.$$

A detailed discussion of these equations can be found in [1], [4], [15], and [16].

Just as we generalized the Verhulst equation by replacing $\{1 - (N/\theta)\}$ by $\{1 - (N/\theta)^{\alpha}\}/\alpha$, we can also generalize Volterra's equations by replacing the N_j in the collision term by N_j^{α}. Let us consider the nonlinear set of rate equations

$$(3.7) \qquad dN_i/dt = k_i N_i + \beta_i^{-1} \sum_{j=1}^{n} a_{ij} N_i (N_j^{\alpha} - 1)/\alpha,$$

$$i = 1, 2, \cdots, n,$$

with the a_{ij}'s again being antisymmetric so that

$$(3.8) \qquad a_{ij} = -a_{ji}.$$

This set of equations has an alternative form

$$(3.9a) \qquad dN_i/dt = k_i' N_i + \beta_i^{-1} \sum_{j} a_{ij} N_j^{\alpha}/\alpha$$

where

$$(3.9b) \qquad k_i' \equiv k_i - \beta_i^{-1} \sum_{j} a_{ij}/\alpha.$$

Volterra's equations are of course exactly of the form (3.9a) with $\alpha = 1$.

If the Volterra collision term, which is proportional to $N_i N_j$ is identified as a standard, then $\alpha < 1$ represents a situation in which the prey adapt themselves somewhat to the growing menace of their predators so that they are affected to a lesser degree than they would have been in the standard Volterra case. The regime $\alpha > 1$ represents a situation in which the prey become exhausted by the predator so that they suffer to a greater degree from the increase in predator population than they would have in the Volterra case.

The interaction model defined by (3.7) enjoys a constant of the motion for a general α in the same manner that is known for the Volterra $\alpha = 1$ case. We first note that

$$\beta_i \, d \log N_i/dt = k_i\beta_i + \sum_{j=1}^{n} a_{ij}(N_j^\alpha - 1)/\alpha,$$

(3.10)
$$i = 1, 2, \cdots, n.$$

The steady state solution of these equations corresponds to $dN_i/dt = 0$ for all i. Let $\{q_i\}$ be the set of steady state solutions. Then the numbers

(3.11)
$$f_j \equiv (q_j^\alpha - 1)/\alpha$$

are solutions of a set of linear equations whose coefficients form an antisymmetrical matrix (3.8). Since

(3.12)
$$k_i\beta_i = -\sum_j a_{ij}(q_j^\alpha - 1)/\alpha,$$

equation (3.10) becomes

(3.13)
$$\beta_i \, d \log N_i/dt = \sum_{j=1}^{n} a_{ij}(N_j^\alpha - q_j^\alpha)/\alpha.$$

Let us multiply both sides of this equation by $(N_i^\alpha - q_i^\alpha)/\alpha$ and sum over all i. Then, in view of the antisymmetry of the a_{ij}'s, the right-hand side of the resulting equation vanishes so that

(3.14)
$$\sum_i \beta_i \alpha^{-1} N_i^{-1}(N_i^\alpha - q_i^\alpha) \, dN_i/dt = 0$$

or

(3.15)
$$(d/dt)\sum_i \alpha^{-1}\beta_i q_i^\alpha \{\alpha^{-1}(N_i/q_i)^\alpha - \log(N_i/q_i)\} = 0.$$

Hence, if we let

(3.16)
$$v_i = \log(N_i/q_i) \quad \text{or} \quad (N_i/q_i) = \exp v_i,$$

we see that (since $\sum \beta_i q_i^\alpha \alpha^{-2}$ is a constant)

(3.17)
$$\sum_i \beta_i q_i^\alpha \{\alpha^{-2}e^{\alpha v_i} - v_i\alpha^{-1} - \alpha^{-2}\} = G = \text{constant}.$$

When we choose $\alpha = 1$, we recover Volterra's constant

of the motion

(3.18) $G_1 = \sum_i \beta_i q_i(-1 - v_i + \exp v_i)$.

When α is chosen to be zero, G_0 is a quadratic form in v_i analogous to harmonic oscillators in physics,

(3.19) $G_0 = \dfrac{1}{2} \sum_1^n \beta_i v_i^2$.

The equations of motion (3.13) appropriate for the $\alpha = 0$ case are linear in the variable v_i:

(3.20) $\beta_i dv_i/dt = \sum_{n=1}^n a_{ij} v_j, \qquad i = 1, 2, \cdots, n$.

To find the solution of the linear equations (3.20), we proceed as follows. We define the set of numbers V_i to be the equilibrium solutions of (3.12) so that

(3.21) $V_i = (q_i^\alpha - 1)/\alpha$.

Since the a_{ij}'s are antisymmetrical, the solutions of the set can be constructed in terms of Pfaffians rather than determinants providing, as is shown in [1], a simpler algorithm for the calculation of the V_j's than the usual ratio of two determinants. Once the V_j's are known, the q_j's follow from (3.21).

It is easy to show (since a_{ij} is antisymmetric) that [1], if $x_i = \beta_i^{1/2} v_i$,

(3.23) $x_i(t) = \sum_{lj} A_{il} A_{jl}^* x_j(0) \exp t\lambda_l$

where the λ_l's are the characteristic values of the matrix (b_{ij}) and the A_{il}'s are components of its characteristic vectors; where $b_{ij} = (a_{ij}/[\beta_i \beta_j]^{1/2})$,

(3.24) $\sum_k b_{ik} A_{kl} = \lambda_l A_i \quad \text{and} \quad \sum_l A_{li}^* A_{lj} = \delta_{ij}$.

The λ_l's are also purely imaginary. Alternatively,

(3.25) $$x_i(t) = \sum M_{ij}(t) x_j(0)$$

where

(3.26) $$M_{ij}(t) = \sum_l A_{il} A_{jl}^* \exp t\lambda_\alpha.$$

It can also be shown that [1]

(3.27) $$\sum_j M_{ij}(t) M_{kj}(t) = \delta_{ik}.$$

Since the characteristic values $\lambda_m = i\omega_m$ and $-i\omega_m$ both appear in (3.26), one can write

(3.28) $$M_{ii}(t) = 2 \sum_{\alpha=1}^N |A_{i\alpha}|^2 \cos t\omega_\alpha, \qquad N = \tfrac{1}{2} n.$$

If all the ω_α are distinct, $M_{ii}(t)$ is almost periodic so that, if it achieves any value once, it will achieve it an infinite number of times.

In the special case $n = 2$, there is only one normal mode frequency so that the time variations of $v_1(t)$ and $v_2(t)$ are purely sinusoidal. The basic equations are

(3.29) $$\beta_1 \dot{v}_1 = -a_{21} v_2 \quad \text{and} \quad \beta_2 \dot{v}_2 = a_{21} v_1$$

where we choose $a_{21} > 0$. The constant of the motion is

(3.30) $$\beta_1 v_1^2 + \beta_2 v_2^2 = c^2 \equiv \beta_1 v_1^2(0) + \beta_2 v_2^2(0).$$

If we let

(3.31) $$\tau = a_{21} t,$$

it is easy to verify that the solutions of (2.39) are

(3.32a) $$v_1 = \alpha \beta_2^{1/2} \cos(\omega\tau + \delta),$$

(3.32b) $$v_2 = \alpha \beta_1^{1/2} \sin(\omega\tau + \delta),$$

where

(3.33) $$\omega = (\beta_1 \beta_2)^{-1/2}; \qquad \kappa^2 = c^2 / (\beta_1^2 + \beta_2^2)$$

and

(3.34) $$\delta = \cos^{-1}\{v_1(0)/\kappa\beta^{1/2}\}.$$

The solutions of the nonlinear equations are

(3.35a) $$N_1(t) = q_1\exp\{\kappa\beta_2^{1/2}\cos(\omega ta_{21} + \delta)\},$$

(3.35b) $$N_2(t) = q_2\exp\{\kappa\beta_1^{1/2}\sin(\omega ta_{21} + \delta)\},$$

where

(3.36) $$q_1 = \exp(a_{12}\beta_2 k_2); \qquad q_2 = \exp(-a_{12}\beta_1 k_1).$$

Since we chose $a_{12} > 0$, we must also choose $k_1 < 0$ and $k_2 > 0$ if we wish species 1 to grow exponentially in the absence of species 2, and species 2 to die out in the absence of 1's.

An important feature of solutions (2.45) is that as κ becomes large, the peaks in N_1 and N_2 become sharper and more distorted from sine waves. From (2.43) and (2.40), it is clear that large initial deviations from equilibrium correspond to large κ and thus to sharp peaks of the type observed in Figure 5 (and also obtained by machine calculation of the solutions of the Lotka-Volterra equations [1], [17]).

It is easy to discuss three, four, and five-species populations in a similar way. This will be presented elsewhere. The advantage of our model over the Lotka-Volterra one is that such calculations are elementary in our model while elaborate computer calculations must be made to investigate the Lotka-Volterra models with more than two species. From all we have been able to ascertain, the nonlinear qualitative features of both are similar. Also, since there is no assurance that the Volterra model represents nature any better than various others, there is no use for one to torment himself with it for developing an intuition about interacting species.

Once we become accustomed to the freedom of changing

the form of the collision or interaction term in (3.7), there is no reason why we must keep the N_i's to the first power in the equation. We could just as well consider a set of models whose rate equations are

(3.37)
$$dN_i/dt = k_i N_i^\gamma + \beta_i^{-1} \sum_{j=1}^n a_{ij} N_i^\gamma (N_j^\alpha - 1)/\alpha,$$
$$i = 1, \cdots, n,$$

so that

(3.38a) $$\beta_i (1 - \gamma)^{-1} dN_i^{1-\gamma}/dt = k_i \beta_i + \sum_{j=1}^n a_{ij} (N_j^\alpha - 1)/\alpha$$

(3.38b) $$= \sum_{i=1}^n a_{ij} (N_i^\alpha - q_i^\alpha)/\alpha$$

or better,

(3.39) $$(1 - \gamma)^{-1} \beta_i d(N_i^{1-\gamma} - q_i^{1-\gamma})/dt = \sum_{i=1}^n a_{ij} (N_i^\alpha - q_i^\alpha)/\alpha.$$

We can immediately construct a class of "solvable" models by choosing $(1 - \gamma) = \alpha$ so that if we let

(3.40) $$u_i = (N_i^\alpha - q_i^\alpha)/\alpha \quad \text{or} \quad N_i = q_i (1 + \alpha u_i q_i^{-\alpha})^{1/\alpha},$$

(3.41) $$\beta_i \frac{du_i}{dt} = \sum_{j=1}^n a_{ij} u_j,$$

which is a set of linear equations and, therefore, solvable by standard methods. The choice $\alpha = 0$ is, of course, the one considered earlier.

There is one choice of γ which gives a symmetry in the binary collision term, namely $\gamma = \frac{1}{2}$ for then

(3.42) $$\frac{dN_i}{dt} = k_i N_i^{1/2} + 2\beta_i^{-1} \sum_{j=1}^n a_{ij} N_i^{1/2} (N_j^{1/2} - 1).$$

This is linearized by the transformation

(3.43) $$u_i = 2(N_i^{1/2} - q_i^{1/2}).$$

This also suggests a broader class of approaches to saturation for a single species

$$(3.44) \qquad dN/dt = kN^\gamma [1 - (N/\theta)^\alpha]/\alpha$$

which, in the case of $\gamma = \alpha = \frac{1}{2}$, leads to

$$(3.45) \qquad N(t) = \theta \{ 1 - [1 - (N_0/\theta)^{1/2}] \exp(- kt/\theta^{1/2}) \}^2,$$

which might also be used to fit population growth data.

The set of equations (3.38) also has a conservation law when a_{ij} is antisymmetric. One finds in the same manner as was used to derive (3.17) that

$$(3.46) \qquad G_{\alpha,\gamma} = \sum_i \frac{\beta_i q_i^{\alpha+1-\gamma}}{\alpha(1-\gamma)} \left\{ \left(\frac{1-\gamma}{1+\alpha-\gamma} \right) \left(\frac{N_i}{q_i} \right)^{1+\alpha-\gamma} - \left(\frac{N_i}{q_i} \right)^{1-\gamma} \right\} = \text{constant}$$

or, in terms of v_i (see (3.16)),

$$(3.47) \qquad G_{\alpha,\gamma} = \sum_i \frac{\beta_i q_i^{\alpha+1-\gamma}}{\alpha(1-\gamma)} \left\{ 1 + \left(\frac{1-\gamma}{1+\alpha-\gamma} \right) \right.$$
$$\left. \cdot [- 1 + \exp v_i(1+\alpha-\gamma)] - \exp v_i(1-\gamma) \right\}$$
$$= \text{constant}.$$

We see that $G_{\alpha,1}$ is exactly (3.17).

An interesting feature of the generalization of the Volterra equation (3.7) is that in a system not far from equilibrium (but still in the nonlinear regime) it can be approximated by an equivalent Volterra system. We use the form (3.13). Then

$$(3.48) \qquad \beta_i d \log N_i/dt = \sum_{j=1}^n a_{ij} \alpha^{-1}(N_j - q_j) \left\{ \frac{N_j^\alpha - q_j^\alpha}{N_j - q_j} \right\}.$$

But if $x \equiv N/q$

$$\frac{N^\alpha - q^\alpha}{N - q} = q^{\alpha-1}\left(\frac{x^\alpha - 1}{x - 1}\right) = q^{\alpha-1}\left\{\frac{[1 + (x - 1)]^\alpha - 1}{x - 1}\right\}$$

$$(3.49) \qquad = \alpha q^{\alpha-1}\{1 + \tfrac{1}{2}(\alpha - 1)(x - 1)$$
$$+ \tfrac{1}{6}(\alpha - 1)(\alpha - 2)(x - 1)^2 + \cdots\}.$$

Hence (3.48) becomes

$$\beta_i d\log(N_i/q_i)/dt$$

$$= \sum_{j=1}^{n} a_{ij} q_j^{\alpha-1}(N_j - q_j)\left\{1 + \frac{1}{2}(\alpha - 1)\left(\frac{N_j}{q_j} - 1\right)\right.$$

$$(3.50) \qquad\qquad \left. + \frac{1}{6}(\alpha - 1)(\alpha - 2)\left(\frac{N_j}{q_j} - 1\right)^2 + \cdots\right\}.$$

If $[(N_j/q_j) - 1]$ is sufficiently small the higher order terms in the bracket can be neglected.

The almost equivalent Volterra system can be found as follows. Rewrite (3.50) as

$$\beta_i d\log(N_i/q_i)/dt \cong -\sum_{j=1}^{n} a_{ij} q_j^\alpha + \sum_j a_{ij} q_j^{\alpha-1} N_j$$

or

$$\beta_i q_i^{\alpha-1} d\log(N_i/q_i)\, dt \cong -q_i^{-1}\sum_{j=1}^{n} a_{ij} q_i^\alpha q_j^\alpha$$

$$(3.51) \qquad\qquad + \sum_j a_{ij} q_i^{\alpha-1} q_j^{\alpha-1} N_j.$$

Then define a new set of antisymmetric matrix elements

$$(3.52) \qquad\qquad A_{ij} \equiv (q_i q_j)^{\alpha-1} a_{ij},$$

a set of equivalence numbers,

$$(3.53) \qquad\qquad B_i \equiv \beta_i q_i^{\alpha-1},$$

and a set of K_i such that

$$(3.54) \qquad\qquad B_i K_i = -q_i^{-1}\sum_j a_{ij} q_i^\alpha q_i^\alpha.$$

Then, (3.50) becomes the Volterra set

$$(3.55) \qquad B_i d \log N_i/dt = B_i K_i + \sum_j A_{ij} N_j.$$

One can also proceed in the opposite direction. Let us start with a Volterra set

$$(3.56) \qquad B_i d \log N_i/dt = B_i K_i + \sum_j A_{ij} N_j$$

where the $\{B_i\}$, $\{K_i\}$ and $\{A_{ij}\}$ are postulated to be known. We wish to find a set of B_i's, K_i's and a_{ij}'s such that

$$(3.57) \qquad \beta_i d \log N_i/dt = \beta_i K_i + \sum a_{ij}(N_j^\alpha - 1)/\alpha$$

gives a first order fit to the Volterra equation. First define $\{q_i\}$ such that $(q_i^\alpha - 1)/\alpha \equiv Q_i$, the equilibrium solution of (3.56). With the q_i's known, define $a_{ij} \equiv A_{ij}(q_i q_j)^{1-\alpha}$. The β_i and k_i must now be defined. Rewrite (3.57) in the approximate form (see (3.50))

$$(3.58) \qquad \beta_i q_i^{\alpha-1} d \log N_i/dt = \sum_j A_{ij}(N_j - q_j).$$

Then define $\beta_i \equiv B_i q_i^{1-\alpha}$. With β_i known we next set

$$(3.59) \qquad \beta_i k_i q_i^{\alpha-1} = -\sum_j A_{ij} q_j,$$

thus defining the k_i so that all required quantities have been evaluated.

4. Statistical theory of interacting species in presence of random disturbances

There are two approaches to nonequilibrium statistical mechanics. The traditional one is the Brownian motion type theory in which an assembly of particles is divided into two sets, one being a small set of particles of interest and the other a set of an enormous number of particles whose influence is characterized by a random force. The second approach is the more modern (and more complicated) one in which one uses the laws of dynamics at all times, but in which a certain lack of information exists in the initial conditions of the problem. It is assumed that information is given about the initial values of a small number of variables while only statistical statements are made about the remainder of them. The equations of motion (with no random forces or averaging) are solved as a function of the time so that the only way statistics enters the problem is through the initial conditions. We use both of these techniques to develop statistical theories of interacting species. In this section we employ the techniques of Brownian motion theory and, in the next, all statistics will be put into the initial conditions. For simplicity we will only discuss our solvable models whose basic equations of motion are

$$\beta_i \, dN_i/dt = k_i \beta_i N_i^{1-\alpha} + N_i^{1-\alpha} \sum_{j=1}^{n} a_{ij}(N_j^{\alpha} - 1)/\alpha,$$

(4.1) $$i = 1, 2, \cdots, n,$$

$$a_{ij} = -a_{ji}.$$

We now assume that many influences are not included in this equation. They may be the influence of a myriad of other species that we do not wish to take into account in detail, changes in weather, etc. All these influences are to be incorporated into a random term of the form $N_i^{1-\alpha} F_i(t)$ so that

(4.2)
$$\beta_i \, dN_i/dt = k_i \beta_i N_i^{1-\alpha} + N_i^{1-\alpha} \sum_j a_{ij}(N_j^{\alpha} - 1)/\alpha$$
$$+ N_i^{1-\alpha} F_i(t)$$

so that

(4.3) $$\alpha^{-1}\beta_j \, d(N_i^{\alpha} - 1)/dt = k_i \beta_i + \sum_j a_{ij}(N_j^{\alpha} - 1)/\alpha + F_i(t).$$

Let us *define* a set of numbers $\{q_j\}$ such that

(4.4) $$k_i \beta_i + \sum_j a_{ij} (q_i^\alpha - 1)/\alpha = 0.$$

Then

(4.5) $$\beta_i d\big\{ (N_i^\alpha - q_i^\alpha)/\alpha \big\}/dt = \sum_j a_{ij} (N_j^\alpha - q_j^\alpha)/\alpha + F_i(t).$$

As in §1, we postulate the random function $F_i(t)$ to be generated by a Gaussian random process such that

(4.6a) $$\langle F_i(t) \rangle = 0,$$

(4.6b) $$\langle F_i(t_1) F_j(t_2) \rangle = \sigma_{ij}^2 \delta(t_1 - t_2).$$

Some correlation between species is permitted because some random forces such as weather fluctuations may affect all species simultaneously.

We introduce a new variable

(4.7) $$u_i = (N_i^\alpha - q_i^\alpha)/\alpha.$$

Then

(4.8) $$\beta_i du_i/dt = \sum_j a_{ij} u_j + F_i(t).$$

As is shown in Appendix I, this equation is also applicable when a saturation-inducing factor

(4.9) $$\big\{ 1 - (N_i/\theta_i)^\alpha \big\}$$

is introduced for those k_i with $k_i > 0$. For those i's the first term in (4.2) is

(4.10) $$k_i \beta_i N_i^{1-\alpha} \big\{ 1 - (N_i/\theta_i)^\alpha \big\}.$$

The standard Fokker-Planck equation [12] for the probability that $u_1, u_2, \cdots, u_n$ have preassigned values at a given time t, $P(u_1, u_2, \cdots, u_n; t)$ is

(4.11) $$\frac{\partial P}{\partial t} = - \sum_{i=1}^n \frac{\partial}{\partial u_i} (A_i P) + \frac{1}{2} \sum_{ij} \frac{\partial^2}{\partial u_i \partial u_j} [B_{ij} P]$$

where

(4.12a) $\quad A_i = \lim\limits_{\Delta t \to \theta} \Delta u_i / \Delta t = \beta_i^{-1} \sum\limits_j a_{ij} u_j,$

(4.12b) $\quad B_{ij} = \lim\limits_{\Delta t \to \theta} (\Delta u_i \Delta u_j) / \Delta t = (\beta_i \beta_j)^{-1} \sigma_{ij}^2.$

Equation (3.11) is just the form that has been discussed by Wang and Uhlenbeck [12] in the theory of Brownian motion of coupled oscillators. Following the ideas used in [12], we introduce a new set of variables

(4.13) $\quad z_j = \sum\limits_k c_{jk} u_k \quad \text{with} \quad u_k = \sum\limits_m c_{km}^{(-1)} z_m.$

The matrix C whose elements are c_{jk} and whose inverse C^{-1} has elements $c_{km}^{(-1)}$ is chosen to be that matrix which diagonalizes a matrix A whose elements are (a_{ij}/β_i); i.e., $CAC^{-1} = I$. Since

(4.14) $\quad \partial / \partial u_i \equiv \sum\limits_l (\partial z_l / \partial u_i) \partial / \partial z_l \equiv \sum\limits_l c_{li} \partial / \partial z_l$

and

(4.15) $\quad \sum\limits_{ij} c_{li} (a_{ij}/\beta_i) c_{jm}^{(-1)} = \delta_{l,m},$

we find

(4.16) $\quad \begin{aligned} \sum\limits_j \frac{\partial}{\partial u_i} (A_i P) &= \sum\limits_{i,j} \frac{\partial}{\partial u_i} [(a_{ij}/\beta_i) u_j P] \\ &= \sum\limits_l \lambda_l \partial (P z_l) / \partial z_l, \end{aligned}$

where λ_l is the eigenvalue of

(4.17) $\quad \sum\limits_i c_{li} [a_{ij}/\beta_i] = \lambda_l c_{lj}.$

On this basis our basic equation is transformed to

(4.18) $\quad \dfrac{\partial P}{\partial t} = -\sum\limits_{l=1}^{n} \lambda_l \dfrac{\partial}{\partial z_l} (P z_l) + \dfrac{1}{2} \sum\limits_{ij} f_{ij} \dfrac{\partial^2 P}{\partial z_i \partial z_j}$

where the f_{ij} are matrix elements of $(CS\widetilde{C})$ where S is the matrix whose elements are $(\sigma_{ij}^2/\beta_j) \equiv s_{ij}$. $\widetilde{C}$ is the transpose of C.

This equation has been solved by Wang and Uhlenbeck for the initial conditions

(4.19a)
$$P(z_1, z_2, \cdots, z_n; 0)$$
$$= \delta(z_1 - z_{10})\delta(z_2 - z_{20}) \cdots \delta(z_1 - z_{10}).$$

By introducing the Fourier transform $f(\xi_1, \cdots, \xi_n; t)$,

$$f(\xi_1, \cdots, \xi_n; t)$$

(4.19b)
$$= \int \cdots \int_{-\infty}^{\infty} dz_1 \cdots dz_n P(z_1 \cdots z_n; t) \exp - i\xi \cdot z,$$

the differential equation (3.18) is reduced to one of the first order which can be solved by the method of characteristics. It is found that [12]

(4.20)
$$f(\xi; t) = \exp\left[-i\sum_j \xi_j z_{j0} \exp \lambda_j t \right.$$
$$\left. + \sum_{ij} f_{ij} \left\{ \frac{\xi_i \xi_j}{\lambda_i + \lambda_j} \right\} \{1 - \exp(\lambda_i + \lambda_j)t\} \right].$$

This is the Fourier transform of an n-dimensional Gaussian distribution with the average values

(4.21a)
$$\langle z_i \rangle = z_{i0} \exp(\lambda_i t) \equiv \bar{z}_i$$

and the variances

(4.21b)
$$\langle (z_i - \bar{z}_i)(z_j - \bar{z}_j) \rangle$$
$$= -\frac{f_{ij}}{(\lambda_i + \lambda_j)} [1 - \exp(\lambda_i + \lambda_j)t].$$

The λ_i's are purely imaginary when no saturation effects exist in the original rate equations, i.e., when $a_{ij} = -a_{ji}$. If we rewrite (4.17) as

(4.22a)
$$\sum_i [c_{li}/(\beta_l \beta_i)^{1/2}](a_{ij}/(\beta_i \beta_j)^{1/2}) = \lambda_l [c_{lj}/(\beta_l \beta_j)^{1/2}],$$

then the antisymmetry

(4.22b)
$$a_{ij}/(\beta_i \beta_j)^{1/2} = -a_{ij}/(\beta_i \beta_j)^{1/2}$$

implies that the λ_i's are purely imaginary. The u_i's which are a linear combination of the z_i's then oscillate around their mean values, as do the populations N_i which are related to the u_i's through (4.13).

Our required statistical problem is solved in principle because the distribution function of the z_i's is obtainable from (4.10). This can be transformed into a distribution of the u_i's by using (4.13) and the resulting distribution of the u_i's is converted into one on the N_i's through (4.7). We will discuss these transformations in detail and apply them to some special problems elsewhere. It is not difficult to obtain the Fokker-Planck equation from the analogue of (4.2) even when the exponents on N_i and N_j are not chosen in such a manner as to yield an exactly solvable model. However, since we have not made any interesting deductions from it, we do not present it here.

5. Statistical theory of rate equations of a closed system

We now return to our solvable models of interacting species as characterized by the rate equations (3.1)

$$(5.1) \qquad \beta_i \, dN_i/dt = k_i \beta_i N_i^{1-\alpha} + N_i^{1-\alpha} \sum_j a_{ij}(N_j^\alpha - 1)/\alpha$$

without random influences. Statistics will enter our analysis through ignorance of the initial conditions. We remember the constant of the motion (3.46) with $\gamma = 1 - \alpha$,

$$(5.2) \qquad G_{\alpha,1-\alpha} = \sum_i \beta_i q^{2\alpha} \{\tfrac{1}{2} (N_i/q_i)^{2\alpha} - (N_i/q_i)^\alpha\}/\alpha^2,$$

or

$$(5.3) \qquad U_\alpha = \sum_i \tfrac{1}{2} \beta_i q^{2\alpha} \{ (N_i/q_i)^\alpha - 1\}/\alpha^2 = \text{constant},$$

$$(5.4) \qquad U_\alpha = \tfrac{1}{2} \sum_i \beta_i u_i^2 = \text{constant}$$

where u_i is given by (3.40)

(5.5) $$u_i = (N_i^\alpha - q_i^\alpha)/\alpha.$$

The equations of motion (3.41) are linear in the variable

(5.6) $$x_i = u_i \beta_i^{-1/2}$$

so that

(5.7) $$dx_i/dt = \sum_{j=1}^{n} b_{ij} x_j \quad \text{with} \quad b_{ij} = a_{ij}/(\beta_i \beta_j)^{1/2}$$

and

(5.8) $$U_\alpha = \tfrac{1}{2} \sum x_j^2 = \text{constant.}$$

The solution of (5.5) as given by (3.23) is

(5.9) $$x_i(t) = \sum_{lj} A_{il} A_{ji}^* x_i(0) \exp t\lambda_l \equiv \sum_j M_{ij}(t) x_j(0)$$

with $M_{ij}(t)$ given by (3.26).

We proceed now in the manner followed in [18] in the theory of lattice vibrations and in [1] in the analysis of the Volterra model in the regime of small deviations from equilibrium. Here the results are exact because no assumption of small deviations has been made for our model. We seek the probability distribution of $x_i(t)$ for a fixed i as a function of time when (a) it is known that at time $t = 0$ that x_i has the value $x_i(0)$ and (b) equal probability is assigned to every initial distribution of the x_j's which is consistent with the constant of motion (5.4). Equation (5.9) can be rewritten as

(5.10) $$Y(t) \equiv x_i(t) - M_{ii}(t) x_i(0) = \sum_j' M_{ij}(t) x_j(0).$$

Our problem then is to investigate the statistical properties of $Y_i(t)$ when at time $t = 0$ all sets of $\{x_i(0)\}$ (with $j \neq i$) which satisfy

(5.11) $$R^2 = \sum_j' x_j^2(0) = 2U_\alpha - x_i^2(0)$$

are given equal weight. This problem has been solved by Mazur and Montroll [18] in the context of vibrations of crystal lattices. When n is large, the probability distribution of $Y(t)$ is Gaussian (a discussion for all positive integral n, large or small, can be found in [19])

$$f[Y(t)] \equiv g[x_i(t)\,|\,x_i(0)\,] = \{\,(n/2\pi)^{1/2}/\sigma R\,\}$$

$$(5.12) \qquad\qquad \cdot \exp\{-n[x_i(t) - M_{ii}(t)\,x_i(0)\,]^2/2R^2\sigma^2\}$$

where

$$(5.13) \qquad\qquad \sigma^2 = \sum_{j \neq i} M_{ij}^2(t)\,.$$

We notice from (3.27) that $\sum_j M_{ij}^2(t) = 1$. Therefore

$$(5.14) \qquad\qquad \sigma^2 = 1 - M_{ii}^2(t)$$

where M_{ii} is the almost periodic function (when all λ_l's are distinct) given by (3.28).

In view of the normalization of the $A_{i\alpha}$, (3.24), each $A_{i\alpha}$ is of $O(N^{-1/2})$ so that $|A_{i\alpha}|^2 = O(N^{-1})$. Since $M_{ii}(0) = 1$, $\sigma^2(0) = 1$. Now let us write (3.28) as

$$(5.15) \qquad M_{ii}(t) = (2/N) \sum_{\alpha} |A_{i\alpha}N^{1/2}|^2 \cos t\omega_\alpha.$$

If at a given time the various cosines are completely out of phase so that there are roughly as many positive as negative ones, one might consider the sum to be that of N independent random variables, each of $O(1)$. Then by the central limit theorem for the sum of independent random variables, the expected value of the sum would be of $O(N^{1/2})$ so that $M_{ii}(t) = O(N^{-1/2})$ and, as the number of interacting species becomes large, $M_{ii}(t)$ would become small so that the term $M_{ii}(t)\,x_i(0)$ could be neglected in (4.15) and $\sigma = 1$. In that case $Y(t)$ would have a Gaussian distribution which is independent of $x_i(0)$ as $t \to \infty$.

(5.16)
$$g[x_i(t) \mid x_i(0)]$$
$$\rightarrow (2\pi R^2/n)^{-1/2} \exp\{ -[x_i(t)]^2/2(R^2/n) \}.$$

As $n \rightarrow \infty$, $x_i^2(0)$ can be neglected compared with $2U_\alpha$ and R^2/n is just $(1/n)$th part of the constant of the motion $2U_\alpha$ independently of the species i. The distribution function (5.16) is an analogue of the Maxwell distribution of velocity in statistical mechanics and U_α/n is the analogue of the mean kinetic energy per gas particle. Hence (5.16) is also the analogue of the canonical ensemble distribution function of statistical mechanics as has been conjectured by Kerner for the Volterra model.

The above heuristic remarks can be put on a more rigorous basis by phrasing the discussion of $M_{ii}(t)$ in a somewhat different manner. We might ask how frequently $M_{ii}(t)$ achieves a preassigned value, a. There are two regimes of interest, those values of a which are of $O(N^{-1/2})$ and those which are of $O(1)$. The following theorem [19], [20] is important in the first (or the "noise" regime): Let $L(a)$ be the mean frequency with which the sum

(5.17)
$$M(t) = \frac{1}{n} \sum_1^n a_j \cos t\omega_j$$

achieves the value a. If, as $n \rightarrow \infty$,

(5.18) $n^{-2} \sum a_j^4 \rightarrow 0, \quad n^{-2} \sum \omega_j^4 \rightarrow 0, \quad n^{-2} \sum a_j^4 \omega_j^2 \rightarrow 0,$

and if

(5.19) $\omega_0^2 = \dfrac{1}{n} \sum a_j^2 \omega_j^2 \quad \text{with} \quad a^2 = \dfrac{1}{n} \sum a_j^2,$

and if the frequencies ω_j are linearly independent, as $n \rightarrow \infty$,

(5.20) $L(bn^{-1/2}) \sim (\omega_0/\pi a) \exp - (b/a)^2.$

There is a second theorem due to L. B. Slater [21] which is central to a discussion of the second (large fluctuation) regime:

$$L(q) \sim \frac{1}{\pi \Gamma(\frac{1}{2}n + \frac{1}{2})} \left\{ \frac{n^{-1}\sum a_j - q}{2\pi} \right\}^{(n-1)/2} \left\{ \frac{\sum \omega_j^2 |a_j|}{|a_1 a_2 - a_n|} \right\}^{1/2}$$
(5.21)

where $q = O(1)$ and $0 < q < n^{-1}\sum a_j$. In our case, $\sum a_j = \frac{1}{2}n$ because of the normalization of the $\{A_{il}\}$.

From (5.20) the mean frequency with which a value of $M_{ii}(t)$ in the noise range $|M_{ii}| = bn^{-1/2}$ is repeated is of the order of a mean normal mode frequency ω_0. On the other hand, if one is out of the noise range, say with $q \cong \frac{1}{2}$ independently of n, the frequency of M_{ii} achieving the value q drops exponentially with the number of species n; i.e. when n is very large, M_{ii} practically never achieves a value $q = O(1)$ but merely keeps fluctuating in a small band of values of width of $O(n^{-1/2})$. In the limit $n \to \infty$ the distribution function of the ith species about its equilibrium value is, from (5.16) and (5.11),

(5.22) $F_j(x)dx = (4\pi U_\alpha/n)^{-1/2}\exp\{-x^2/(4U_\alpha/n)\}dx.$

Note from (5.5) and (5.6) that

(5.23) $x_i = (N_i^\alpha - q_i^\alpha)/\alpha\beta_i^{1/2} = q_i^\alpha\{(N_i/q_i)^\alpha - 1\}/\alpha\beta_i^{1/2}$

so that

(5.24)
$$dx_i = d(N_i/q_i)\,dx_i/d(N_i/q_i)$$
$$= (q_i^\alpha/\beta_i^{1/2})\,(N_i/q_i)^{\alpha-1}d(N_i/q_i).$$

Hence the distribution function of (N_i/q_i) is probability that (N_i/q_i) lies between y and $y + dy$ is

(5.25)
$$P(y)\,dy = (4\pi U_\alpha/n)^{-1/2}(q_i^\alpha/\beta^{1/2})\,y^{\alpha-1}$$
$$\cdot \exp\{-q_i^{2\alpha}(y^\alpha - 1)^2/(4\chi^2\beta_i U_\alpha/n)\}dy.$$

Joint probabilities of several populations can be found

in a similar manner. This was done in references for the Volterra model by linearizing the nonlinear equations by expanding around the equilibrium values.

Before closing, we might note that a still broader class of model rate equations than (4.1) can be studied following the lines given above,

$$(5.26) \qquad \beta_i \frac{dN_i}{dt} = \frac{1}{F'(N_i)} \left\{ k_i\beta_i + \sum_{j=1}^{n} a_{ij} F(N_j) \right\}$$

where $F(N)$ is an arbitrary function. It was chosen to be

$$(5.27) \qquad\qquad F(N) = (N^\alpha - 1)/\alpha$$

above.

Since the structure of Cowan's equations for the firing of neurons has a similar structure with

$$(5.28) \qquad\qquad F'(N) = [N(1 - N)]^{-1}$$

one can find a class of models such that $F'(1) = F'(0) = 0$ with our required structure. In Cowan's model N appears where $F(N)$ appears in (5.26) and (5.27) where $F'(N)$ appears, and the resulting equations are not solvable by our methods. We will discuss models of the type (5.21) which resemble Cowan's model elsewhere.

In conclusion the author wishes to thank Drs. Goel and Maitra for many interesting discussions about the rate equations of competition. This research was partially supported by ARPA and was monitored by ONR under Contract No. N00014-67-A-0398-0005.

APPENDIX

Many interacting species with saturation in population

In this appendix we consider our equations of species interaction when the population of some of the species becomes saturated. We limit our discussion to our solvable models.

Let us begin with the inclusion of a saturation term in the $\alpha = 0$ form of (3.10)

$$(A.1) \qquad \beta_i d \log N_i/dt = k_i\beta_i + \sum_{j=1}^{n} a_{ij} \log N_i.$$

Of our n species, we suppose that for the first m ($i = 1, 2, 3, \cdots, m$), $k_i < 0$, $a_{ij} = -a_{ji}$. Then we write

$$(A.2a) \qquad \beta_i d \log N_i/dt = k_i\beta_i + \sum_{j=1}^{n} a_{ij} \log N_i, \quad i = 1, \cdots, m,$$

$$(A.2b) \qquad a_{11} = a_{22} = \cdots = a_{mm} = 0.$$

For the remaining $n - m$ species, we assume that $k_i > 0$ and include a Verhulst type saturation term so that

$$\begin{aligned}
\beta_i d \log N_i/dt &= -k_i\beta_i d \log(N_i/\theta_i) + \sum_{j=1}^{n}{}' a_{ij} \log N_j \\
&= k_i\beta_i \log \theta_i + \sum_{j=1}^{n} a_{ij} \log N_j
\end{aligned}$$

(A.3)

where the prime on the summation indicates that the terms with $i = j$ are omitted and where we define

$$(A.4) \quad a_{ii} = -k_i\beta_i \quad \text{when} \quad i = m+1, m+2, \cdots, n.$$

We define a set of rate constants k_i' by

$$(A.5a) \quad k_i' \equiv k_i \qquad \text{if } i = 1, 2, \cdots, m,$$

$$(A.5b) \quad k_i' = k_i \log \theta_i \quad \text{if } i = m+1, m+2, \cdots, n.$$

Then for all i,

$$(A.6) \qquad \beta_i d \log N_i/dt = \beta_i k_i' + \sum_{j} a_{ij} \log N_j.$$

Furthermore, we define a set of numbers q_j such that

$$(A.7) \qquad \beta_i k_i' + \sum q_{ij} \log q_j = 0.$$

Then (A.6) becomes

$$(A.8) \qquad \beta_i d \log(N_i/\theta_i) \, dt = \sum_j a_{ij} \log(N_j/q_j)$$

or, if

$$(A.9) \qquad v_i = \log(N_i/q_i),$$

$$(A.10) \qquad \beta_i \dot{v}_i = \sum_j a_{ij} v_i, \qquad i = 1, 2, \cdots, n,$$

which is the same as the equations derived in §3, equation (3.20), for the case in which no saturation terms are included. The only difference is that we now allow some of the a_{ij} to be nonvanishing.

The effect of the saturation terms is to force the populations of all the species to reach the equilibrium values $\{q_j\}$ instead of oscillating around them. The way this can be seen mathematically is that the characteristic values λ_l in (3.23) have negative real parts when there are nonvanishing negative a_{ii}. Then $v_i(t) \to 0$ as $t \to \infty$ so that from (A.9), $N_i \to q_i$. It is not difficult to make a general proof of this remark. We merely show how the result follows in the two-species case.

Let $n = 2$ and suppose that $a_{22} = -2\epsilon$ and $a_{12} = -a_{21}$. Then, if we let $v_i = a_i \exp \lambda t$, the characteristic determinant for λ is

$$(A.11) \qquad \begin{vmatrix} -\lambda & a_{12}/\beta_1 \\ -a_{12}/\beta_2 & -\lambda_1 - 2(\epsilon/\beta_2) \end{vmatrix} = 0$$

so that

$$(A.12) \qquad \lambda = -(\epsilon/\beta_2) \pm \left\{ (\epsilon/\beta_2)^2 - (a_{12}^2/\beta_1\beta_2) \right\}^{1/2},$$

a quantity whose real part is always negative if $\epsilon/\beta_2 > 0$, $\beta_1, \beta_2 > 0$ and if a_{12} is real.

We can proceed in a similar manner with the more general solvable model whose basic equation without saturation terms is (with $a_{ij} = -a_{ji}$)

(A.13) $\beta_i dN_i/dt = k_i\beta_i N_i^{1-\alpha} + N_i^{1-\alpha}\sum_j a_{ij}(N_j^\alpha - 1)/\alpha.$

With saturation possibilities, we change this equation as was done in the discussion above. We assume that $k_i < 0$ for $i = 1, 2, \cdots, m$ so that we write

(A.14) $\alpha^{-1}\beta_i dN_i^\alpha/dt = k_i\beta_i + \sum_j a_{ij}(N_j^\alpha - 1)/\alpha,$

$$i = 1, 2, \cdots, m,$$

$$a_{11} \equiv a_{22} = \cdots = a_{mm} = 0.$$

When $i = m + 1, m + 2, \cdots, n$ we introduce a saturation inducing term. We assume that $k_i > 0$ if $i > m$ and write

(A.15) $$\alpha^{-1}\beta_i dN_i^\alpha/dt$$
$$= k_i\beta_i[1 - (N_i/\theta_i)^\alpha]/\alpha + \sum_j a_{ij}'(N_j^\alpha - 1)/\alpha$$
$$\text{if } i = m + 1, m + 2, \cdots, n$$

or

(A.16) $\alpha^{-1}\beta_i dN_i^\alpha/dt = k_i\beta_i[1 - \theta_i^{-\alpha}]/\alpha + \sum_j a_{ij}(N_j^\alpha - 1)/\alpha$

where

(A.17) $a_{ii} = -k_i\beta_i/\theta_i^\alpha$ for $i = m + 1, \cdots, n.$

We define a new set of rate constants k_i' by

(A.18a) $k_i' \equiv k_i$ if $i = 1, 2, \cdots, m,$

(A.18b) $k_i' = k_i(1 - \theta_i^{-\alpha})/\alpha$ if $i = m + 1, m + 2, \cdots, n.$

Then for all i

(A.19) $d^{-1}\beta_i dN_i/dt = \beta_i k_i'/\alpha + \sum_j a_{ij}(N_j^\alpha - 1)/\alpha.$

Furthermore we define a set of numbers q_j by

(A.20) $\beta_i k_i'/\alpha + \sum_j a_{ij}(q_j^\alpha - 1)/\alpha,$ $i = 1, \cdots, n.$

Then

$$(A.21) \quad \alpha^{-1}\beta_i d(N_i^\alpha - q_i^\alpha)/dt = \sum_j a_{ij}(N_i^\alpha - q_i^\alpha)/\alpha,$$

so that if we let

$$(A.22) \qquad u_i \equiv (N_i^\alpha - q_i^\alpha)/\alpha,$$

these equations become

$$(A.23) \qquad \beta_i du_i/dt = \sum_{j=1}^n a_{ij} u_j,$$

which is exactly the same form as (3.14) except that some of the a_{ii} do not vanish.

As in the case of the model with the logarithmic nonlinearity, each $N_i \to q_i$ as $t \to \infty$ instead of oscillating around its equilibrium value when a saturation term exists in the set of equations (A.15).

REFERENCES

1. N. S. Goel, S. C. Maitra and E. W. Montroll, Rev. Modern Phys. **43** (1971), 241-276.

2. E. H. Kerner, *Further considerations on the statistical mechanics of biological associations*, Bull. Math. Biophys. **21** (1959), 217-255. MR **21** # 3278.

3. E. W. Montroll, *Lectures in theoretical physics*, University of Colorado, Boulder, Col., XA, 1967, p. 531.

4. V. Volterra, *Leçon de la théorie mathématique de la lutte pour la vie*, Gauthier-Villars, Paris, 1931.

5. T. R. Malthus, *An essay on the principle of population as it affects the future improvement of society*, 1798.

6. P. F. Verhulst, Correspondence Mathématique et Physique **10** (1838), 10; reprinted in *Readings in ecology*, E. J. Kormondy (editor), Prentice-Hall, Englewood Cliffs, N. J., 1965.

7. cf: R. Pearl, *Studies in human biology*, Baltimore, 1924.

8. E. W. Montroll and L. W. Badger, *Introduction to quantitative aspects of social phenomena*, Gordon and Breach, New York, 1971.

9. B. Gompertz, Philos. Trans. Roy. Soc. **115** (1825), 513.

10. *Statistical abstracts of the U. S.*, U. S. Bureau of Census, Washington, D. C., 1968.

11. W. L. Thorp and W. C. Mitchell, *Business annals*, Nat. Bureau of Econ. Res., New York, 1926.

12. M. C. Wang and G. E. Uhlenbeck, *On the theory of the Brownian motion*. II, Rev. Modern Phys. **17** (1945), 323-342. MR **7**, 130.

13. P. M. Morse, Phys. Rev. **34** (1929), 57.

14. C. S. Elton, *Animal ecology*, London, 1927.

15. U. D'Ancona, *The struggle for existence*, Brill, Leiden, 1954.

16. A. J. Lodka, *Elements of mathematical biology*, reprint, Dover, New York, 1956.

17. H. T. Davis, *Introduction to nonlinear differential and integral equations*, Dover, New York, 1962. MR **31** # 6000.

18. P. Mazur and E. Montroll, *Poincaré cycles, ergodicity, and irreversibility in assemblies of coupled harmonic oscillators*, J. Mathematical Phys. **1** (1960), 70-84. MR **26** # 4730.

19. E. W. Montroll, *Lectures in theoretical physics*, University of Colorado, Boulder, Col., III, 1961, p. 221.

20. M. Kac, *On the distribution of values of trigonometric sums with linearly independent frequencies*, Amer. J. Math. **65** (1943), 609-615. MR **5**, 96.

21. L. B. Slater, Proc. Cambridge Philos. Soc. **35** (1939), 56.

AUTHOR INDEX

Roman numbers refer to pages on which a reference is made to a work of an author.

Italic numbers refer to pages on which a complete reference to a work by an author is given.

Boldface numbers indicate the first page of an article in this volume.

SUBJECT INDEX

148